AF388672

Molecular Electronics

Molecular Electronics

Editor

Linda Angela Zotti

MDPI • Basel • Beijing • Wuhan • Barcelona • Belgrade • Manchester • Tokyo • Cluj • Tianjin

Editor
Linda Angela Zotti
Universidad Autónoma de Madrid
Spain

Editorial Office
MDPI
St. Alban-Anlage 66
4052 Basel, Switzerland

This is a reprint of articles from the Special Issue published online in the open access journal *Applied Sciences* (ISSN 2076-3417) (available at: https://www.mdpi.com/journal/applsci/special_issues/Mole_Electronics).

For citation purposes, cite each article independently as indicated on the article page online and as indicated below:

LastName, A.A.; LastName, B.B.; LastName, C.C. Article Title. *Journal Name* **Year**, *Volume Number*, Page Range.

ISBN 978-3-0365-1625-7 (Hbk)
ISBN 978-3-0365-1626-4 (PDF)

Contents

About the Editor

Linda Angela Zotti obtained her MSc in physics from the Universitá degli Studi di Bari (Italy) and her PhD in chemistry from the University of Liverpool (United Kingdom). She then carried out her postdoctoral career in Madrid (Spain) and Dublin (Ireland). She is currently an assistant professor at the Universidad Autónoma de Madrid.

Preface to "Molecular Electronics"

The purpose of this Special Issue is to offer a general overview of the current state of affairs in the field of molecular electronics without the intention of covering every single aspect. We envisage that this Special Issue will be useful for specialists in the field, both theoreticians and experimentalists. Furthermore, given the strong interdisciplinarity of this topic, which lies at the interface between physics, chemistry, biology and engineering, it will be appealing to the non-specialist who is looking to gain a feeling about what is going on at the moment in the field. In times such as these, in which the number of scientific journals and research studies is soaring, Special Issues are a good opportunity for researchers working in the same field to "meet" virtually on a common platform. This has become even more important in the last year due to travel restrictions following the COVID-19 pandemic, which have prevented researchers from meeting in person at international conferences. I am, therefore, very grateful to the authors for their enthusiasm and willingness to take part in the project. Finally, I would like to thank the editorial team of *Applied Sciences* for their invaluable assistance, which has made this a very enjoyable experience.

Linda Angela Zotti
Editor

applied
sciences

MDPI

Editorial

Molecular Electronics

Linda A. Zotti [1,2,3]

1   Departamento de Física Teórica de la Materia Condensada, Universidad Autónoma de Madrid,
    E-28049 Madrid, Spain; linda.zotti@uam.es
2   Condensed Matter Physics Center (IFIMAC), Universidad Autónoma de Madrid, E-28049 Madrid, Spain
3   Departamento de Física Aplicada I, Escuela Politécnica Superior, Universidad de Sevilla,
    E-41011 Seville, Spain

The field of molecular electronics is currently experiencing a renaissance. Recent advances in experimental techniques on the one hand, combined with theoretical modelling on the other, have enabled the exploration of new avenues, as well as allowing for significant progress along more traditional lines. The field is very diverse and embraces many cutting-edge research areas such as quantum interference, thermoelectrics, heat transfer, spintronics, switch devising, and biomolecular electronics. Studies on these subjects are driven by a multitude of needs in modern society: the current global quest for cheap and sustainable technology, the conversion of waste thermal energy, efficient ways of storing and processing information, and the fabrication of biocompatible and implantable devices to name but a few. Last but not least, molecules have proven to be an ideal platform for the advancement of knowledge of fundamental physics. This special issue aims at showcasing some of the many facets of research in molecular electronics and surveying the latest advances in the field. It contains seven studies, including five research papers [1–5] and two reviews [6,7], which span different areas of the field from both a theoretical and experimental perspective.

Funding: L.A.Z. thanks the University of Seville for their financial support through the VI PPIT-US program.

Acknowledgments: This publication was only possible with the invaluable contributions from the authors, reviewers, and the editorial team of Applied Sciences.

Conflicts of Interest: The author declares no conflict of interest.

References

1. Quintans, C.S.; Andrienko, D.; Domke, K.F.; Aravena, D.; Koo, S.; Díez-Pérez, I.; Aragonès, A.C. Tuning Single-Molecule Conductance by Controlled Electric Field-Induced*trans*-to-cis Isomerisation. *Appl. Sci.* **2021**, *11*, 3317. [CrossRef]
2. Romero-Muñiz, C.; Ortega, M.; Vilhena, J.G.; Pérez, R.; Cuevas, J.C.; Zotti, L.A. The Role of Metal Ions in the Electron Transport through Azurin-Based Junctions. *Appl. Sci.* **2021**, *11*, 3732. [CrossRef]
3. Koley, A.; Maiti, S.K.; Ojeda Silva, J.H.; Laroze, D. Spin Dependent Transport through Driven Magnetic System with Aubry-Andre-Harper Modulation. *Appl. Sci.* **2021**, *11*, 2309. [CrossRef]
4. Noori, M.D.; Sangtarash, S.; Sadeghi, H. The Effect of Anchor Group on the Phonon Thermal Conductance of Single Molecule Junctions. *Appl. Sci.* **2021**, *11*, 1066. [CrossRef]
5. Montes, E.; Vázquez, H. Role of the Binding Motifs in the Energy Level Alignment and Conductance of Amine-Gold Linked Molecular Junctions within DFT and DFT +Σ. *Appl. Sci.* **2021**, *11*, 802. [CrossRef]
6. Herrer, L.; Martín, S.; Cea, P. Nanofabrication Techniques in Large-Area Molecular Electronic Devices. *Appl. Sci.* **2020**, *10*, 6064. [CrossRef]
7. Dappe, Y.J. Attenuation Factors in Molecular Electronics: Some Theoretical Concepts. *Appl. Sci.* **2020**, *10*, 6162. [CrossRef]

 check for **updates**

Citation: Zotti, L.A. Molecular Electronics. *Appl. Sci* **2021**, *11*, 4828. https://doi.org/10.3390/app11114828

Received: 18 May 2021
Accepted: 21 May 2021
Published: 25 May 2021

Publisher's Note: MDPI stays neutral with regard to jurisdictional claims in published maps and institutional affiliations.

*applied
sciences*

Article

The Role of Metal Ions in the Electron Transport through Azurin-Based Junctions

Carlos Romero-Muñiz [1,2,3]*, María Ortega [1], Jose Guilherme Vilhena [4], Rubén Pérez [1,5], Juan Carlos Cuevas [1,5] and Linda A. Zotti [1,3,5]*

[1] Departamento de Física Teórica de la Materia Condensada, Universidad Autónoma de Madrid, E-28049 Madrid, Spain; maria.ortega@uam.es (M.O.); ruben.perez@uam.es (R.P.); juancarlos.cuevas@uam.es (J.C.C.)
[2] Department of Physical, Chemical and Natural Systems, Universidad Pablo de Olavide, Ctra. Utrera Km. 1, E-41013 Seville, Spain
[3] Departamento de Física Aplicada I, Escuela Politécnica Superior, Universidad de Sevilla, E-41011, Seville, Spain
[4] Department of Physics, University of Basel, Klingelbergstrasse 82, CH-4056 Basel, Switzerland; guilherme.vilhena@unibas.ch
[5] Condensed Matter Physics Center (IFIMAC), Universidad Autónoma de Madrid, E-28049 Madrid, Spain
* Correspondence: crommun@upo.es (C.R.M.); linda.zotti@uam.es (L.A.Z.)

Abstract: We studied the coherent electron transport through metal–protein–metal junctions based on a blue copper azurin, in which the copper ion was replaced by three different metal ions (Co, Ni and Zn). Our results show that neither the protein structure nor the transmission at the Fermi level change significantly upon metal replacement. The discrepancy with previous experimental observations suggests that the transport mechanism taking place in these types of junctions is probably not fully coherent.

Keywords: azurin; solid-state junction; biomolecular electronics; electronic transport; density functional theory; molecular dynamics

Citation: Romero-Muñiz, C.; Ortega, M.; Vilhena, J.G.; Pérez, R.; Cuevas, J.C.; Zotti, L.A. The Role of Metal Ions in the Electron Transport through Azurin-Based Junctions. *Appl. Sci.* **2021**, *11*, 3732. https://doi.org/10.3390/app11093732

Academic Editor: Cezary Czaplewski

Received: 1 March 2021
Accepted: 16 April 2021
Published: 21 April 2021

Publisher's Note: MDPI stays neutral with regard to jurisdictional claims in published maps and institutional affiliations.

1. Introduction

The study of metal–protein–metal junctions has opened up new avenues in the field of molecular electronics as it paves the way to develop new hybrid devices capable of exploiting proteins' remarkable properties [1]. In particular, they have the inherent capability of transferring charge over long distances. Thus, one can foresee the production of sensors, flexible implants, solar cells and many other types of electrical devices incorporating proteins [2–6]. The significant amount of work accumulated in recent years on this topic has given rise to a new field, namely that of protein-based electronics, which has been given the name of proteotronics [7]. Research in this field has been carried out on various fronts. On the one hand, remarkable effort has been made to understand the exact nature of the charge transfer mechanism in protein-based junctions as well as to identify the components which play the dominant role in the transport process [8–11]. On the other hand, the possibility of modifying the conductance properties by the insertion of mutations [3,10,11] has also attracted a lot of interest. Research on the transport properties of the building blocks of proteins, namely amino acids and peptides, has also been carried out [12–14].

Recent experimental results highlight the possibility of tuning the current density through azurin-based junctions via the replacement of the central copper ion with other metals [15]. These measurements were performed on proteins monolayers lying on a conducting substrate and contacted to a macroscopic electrode at the opposite side. At low temperatures, the difference in current between the Cu(II)-azurin and the Zn-azurin was found to be nearly two orders of magnitude. This was ascribed to the fact that each metal contributes to levels lying at different energies with respect to the Fermi level. Combining organic and metal components in the central area of molecular junctions is indeed a strategy which has been used quite often in molecular electronics [16–18].

In the case of azurins, this possibility becomes particularly intriguing since the aforementioned results of Amdursky et al. [15] indicate that by modifying only one (metal) atom one can modify the conductance properties of a system of almost 2000 atoms. To shed light on this issue, we thus performed theoretical calculations based on density functional theory (DFT) and calculated the electron-transport properties of several metal–azurin–metal junctions based on different metallic ions. Numerous computational studies on the active site on metal-substituted azurins are reported in the literature [19,20], but none involved the study of a whole junction comprising an entire protein held between two metal electrodes. Azurins are one of the proteins which have been studied the most in the field of protein-based electronics; additionally, they have long been the object of many computational studies because of their key role in biological functions [21]. Over the years, various experimental techniques have been adopted to measure the conductance of proteins [6,8,10,22–33]. In this work, we focus on structures which are likely to be formed during the experiments based on STM (Scanning Tunneling Microscopy), which are generally designed to measure the conductance of a single protein. The exact nature of the electron transport through the blue-copper azurin has been the object of a long debate [10,34], the degree of coherence being one of the issues at the heart of the debate. More specifically, what still needs clarifying is whether the process is fully coherent or whether it is better described by a sequential-tunneling process consisting of various steps which are individually coherent. In the case of the metal-substituted proteins studied in [15], the low-temperature results showed evidence for coherent transport; it was suggested that both tunneling and hopping are possible, however, depending on the temperature and the type of junction [35]. For the present study, we are particularly interested in the possibility of fully-coherent transport, which was suggested by the temperature-independence of the conductance measured in several experiments based on azurins [8,11,23,26,32,36–39]. In [34], various types of configurations were studied, in which an STM tip was contacted to a blue-copper azurin in three different ways: in blinking mode (in which the tip is kept at a fixed distance from the surface and the protein is allowed to attach to the tip spontaneously), via top indentation (in which the tip is brought closer to the protein) and by lateral indentation (in which the tip approaches the protein sideways). It was found that, under the assumption of fully-coherent electron transport, measurable conductance values can only be obtained by slightly compressing the protein upon tip indentation from the top or by bringing the tip close to the protein from the side. In this study, we chose to focus on this latter scenario. We compared the conductance properties of the blue-copper azurin from *Pseudomonas aeruginosa* with that of junctions in which the Cu ion was replaced by Zn, Ni or Co. Such a substitution is well known to lead to the formation of stable compounds [40–45], which in fact were used for the aforementioned experiments [15]. The Apo (protein deprived of the metal ion) was also considered.

2. Methods

2.1. Azurin Molecular Dynamics Simulations Free Solved in Water

To inspect structural modifications resulting from the ion-substitution, we performed all-atom molecular dynamics (MD) simulations. To this aim, three different azurin variants were considered: Cu-azurin or native azurin, Co-azurin and Ni-azurin. In all cases, the starting configuration is the one available in the the protein data bank [46,47] (PDBID = 4 AZU), and for each variant the Cu atom was replaced by the corresponding ion. Subsequently, protons were added accordingly to the calculated ionization states [48] of its titratable groups at a pH of 4.5—corresponding to the pH used in reference experiments [10]. In all cases, the resulting structures had a zero net charge. The amino acids inter-atomic interactions were modeled using the standard ff14SB force field [49]. As for the ion-coordination site, composed by the ion and five surrounding amino acids (GLY45, HIS46, CYS112, HIS117 and MET121), we adopted the ab-initio derived force-field parameters reported in [20].

This force-field includes all bonded and non-bonded terms in the coordination sphere for the three metal ions here considered (these parameters are detailed in the Supplementary Materials). Being all simulations performed in an explicit water medium, these solvent atoms were described using the TIP3P force-field [50], which is consistent with the choice of both aforementioned force-field parameters.

All MD simulations were performed using AMBER18 software suite [51] with NVIDIA GPU acceleration [52–54]. Moreover, we used periodic boundary conditions (PBC) with a rectangular box and particle mesh ewald (PME), with a real-space cutoff of 10 Å, to account for long-range electrostatic interactions. Van der Waals (vdW) contacts were truncated at the real space cutoff of 10 Å and coordinates were saved every 1000 steps. The temperature of the system was adjusted by means of a Langevin thermostat [55] with a friction coefficient of $\gamma = 1\ \text{ps}^{-1}$. For the simulations performed in the NPT ensemble, a Berendsen barostat [56] with a relaxation time of $t_p = 1$ ps was used. The SHAKE [57] algorithm was used to constrain bonds containing hydrogen atoms, thus allowing us to use an integration time-step of 2 fs. The simulation protocol is analogous to the one used in [3], which consists of the following four stages: (1) preparing the system by embedding the protein in water in such a way that the minimum solute–water distance is 1 Å, thus resulting in a system with dimensions $\sim$70 Å$\times$63 Å$\times$69 Å; (2) energy minimization of the structures using a combination of steepest descent and conjugate gradient methods; (3) heating up the system from 0 to 300 K with a 2 ns NPT simulation thus ensure an even temperature and density/presure ($T = 300$ K and $P = 1$ atm) on the system; and (4) performing 500 ns-long MD production runs in the NVT ensemble.

2.2. Density Functional Theory Calculations on Metal–Protein–Metal Junctions

We carried out density functional theory (DFT) calculations on metal–protein–metal junctions obtained by replacing, in the geometries obtained for the work in [34], the Cu ion with Zn, Co or Ni without re-optimizing the structure. This strategy makes it possible to identify effects in the electrical conductance due to changes in the electronic structure only, ruling out factors which may rather be related with geometrical differences. As mentioned in the Introduction, we focused on the structures that describe a gold tip approaching the protein sideways since, among the possible scenarios analyzed in [34], this was found to yield one of the highest conductance values. Further details about the protocol followed for obtaining these metal–protein–metal geometries can be found in the Supplementary Materials.

In all cases, the substrate and tip employed in these calculations consist of 969 and 252 atoms, respectively. Note that the corresponding number of gold atoms employed in the MD simulations was instead much larger (2538 and 835, respectively): this is because, when importing the MD output structures for the Cu-based junctions as an input for the DFT calculations, the outermost layers in both the surface and the tip as well as the surrounding water molecules were removed in order to comply with the computational limitations. The OpenMX code was used, which is based on highly optimized numerical pseudoatomic orbitals (PAOs) [58,59] and which we already used for our previous work on proteins [8,34,60,61]. We used the Perdew–Burke–Ernzerhof (PBE) exchange and correlation functional [62] and norm-conserving pseudopotentials [63]. Single-ζ basis sets were used for all species involved in the calculations (including the gold tip and substrate) except for the metallic ion for which a double-ζ basis set was employed instead. The cutoff radii used for the PAOs were 8 Bohr for the metal ion, 7 Bohr for S and 5 Bohr for C, N, O and H. As for the gold electrodes, the Au atoms belonging to the tip and the substrate were described by the cutoff radii of 9 Bohr in those regions closer to the protein, whereas a 7-Bohr cutoff radius was chosen for the rest of the atoms. Due to computational reasons, these calculations were performed without spin polarization. The electronic self-consistency is achieved by a Pulay mixing scheme based on the residual minimization method in the direct inversion iterative subspace (RMM-DIIS) [64] with a Kerker metric [65], using an energy cutoff of 10^{-8} Ha as convergence criterion. Further details about the strategies used to reach the self-consistent-cycle convergence can be found in [60,61]. The overlap and Hamiltonian matrices obtained by the DFT calculations were subsequently used to compute the zero-bias transmission as a function of energy in the spirit of the Landauer formalism and making use of Green's function techniques. Within this method, the structure is divided in three regions (left (L), center (C) and right (R)) and the energy dependent transmission is given by

$$\tau(E) \; = \; \mathrm{Tr}[\mathbf{\Gamma}_\mathrm{L}\mathbf{G}^\mathrm{r}_\mathrm{CC}\mathbf{\Gamma}_\mathrm{R}\mathbf{G}^\mathrm{a}_\mathrm{CC}], \tag{1}$$

where $\mathbf{\Gamma}_\mathrm{L/R}$ are the scattering rate matrices and $\mathbf{G}^\mathrm{r(a)}_\mathrm{CC}$ is the retarded (advanced) Green's function (see [34] for further details).

3. Results and Discussion

First, we inspect the influence of ion replacement on the protein structure. Henceforth, we refer to X-azurin (X = Ni, Cu, Co and Zn) as the azurin containing the X-ion. The four metals considered all belong to the first row of the d block of the periodic table. In Figure 1a–c, we represent the time averaged configuration obtained from a 0.5 µs long MD simulation of the three different proteins, i.e., Cu-azurin (black), Co-azurin (red) and Ni-azurin (green). From the structure overlap between the different proteins and native azurin (i.e., Cu-azurin), it becomes apparent that the ion replacement has little influence on the protein conformation. This is further corroborated and quantified, by both small root-mean-square-deviation (RMSD) between average structures (shown in Figure 1a–c) as well as the preservation of the most prevalent secondary structure on the protein, i.e., β-sheets shown in Figure 1e. The β-sheets refer to a particular spacial organization of the protein's amino acids regulated/characterized by a specific hydrogen bond network among them. In the particular case of azurin, the most prevalent sort of secondary structure is the β-sheet, which are further arranged in a circular manner, thus forming a characteristic β-barrel. In Figure 1e, we quantify the content of this secondary-structure for the different variants to realize that the ion replacement does not alter/disrupt the H-bond network sustaining the characteristic shape and structure of the wild-type azurin. Moreover, from the RMSD, we observed that all proteins exhibit a structural deviation from the native protein smaller than 1.5 Å. Such RMSD values are within the structural variations experienced by the protein resulting from thermal fluctuations. Moreover, the process of protein adsorption and subsequent tip indentation results in a structural deformation which is at least three times larger than the one observed from altering the protein ion (see Figure 2). All considered, one may state that, by altering the coordination ion, the structural modifications thus introduced are substantially smaller than the ones the protein experiences during the contact formation process. Note that the similarity of the parameters governing the bond motif of the different metal ions (e.g., charge and bond distances provided in Figures S2–S5) precluded the negligible structural modifications here observed. All considered, by the aforementioned reasons, the subsequent simulations were performed using a tip-protein contact equilibrated.

Figure 2 shows the initial and final output geometries of the MD simulation performed on the Cu-azurin from which the structures used for the present work were obtained (see more details in [34]). In this simulation, the tip is brought close to the protein sideways and the contact is established via the α-helix arm and hidrophobic patch.

In the left column of Figure 3, we show the zero-bias transmission as a function of energy for the four proteins considered, at four different values of the tip-ion distance (d_{t-X}). Such a distance was evaluated as the distance between the center of the lowest tip layer and the metal ion. In all curves, all resonances appear very sharp, indicating very low hybridization between the corresponding orbitals and the metallic leads. As expected, for most energy values the transmission decreases for increasing distance.

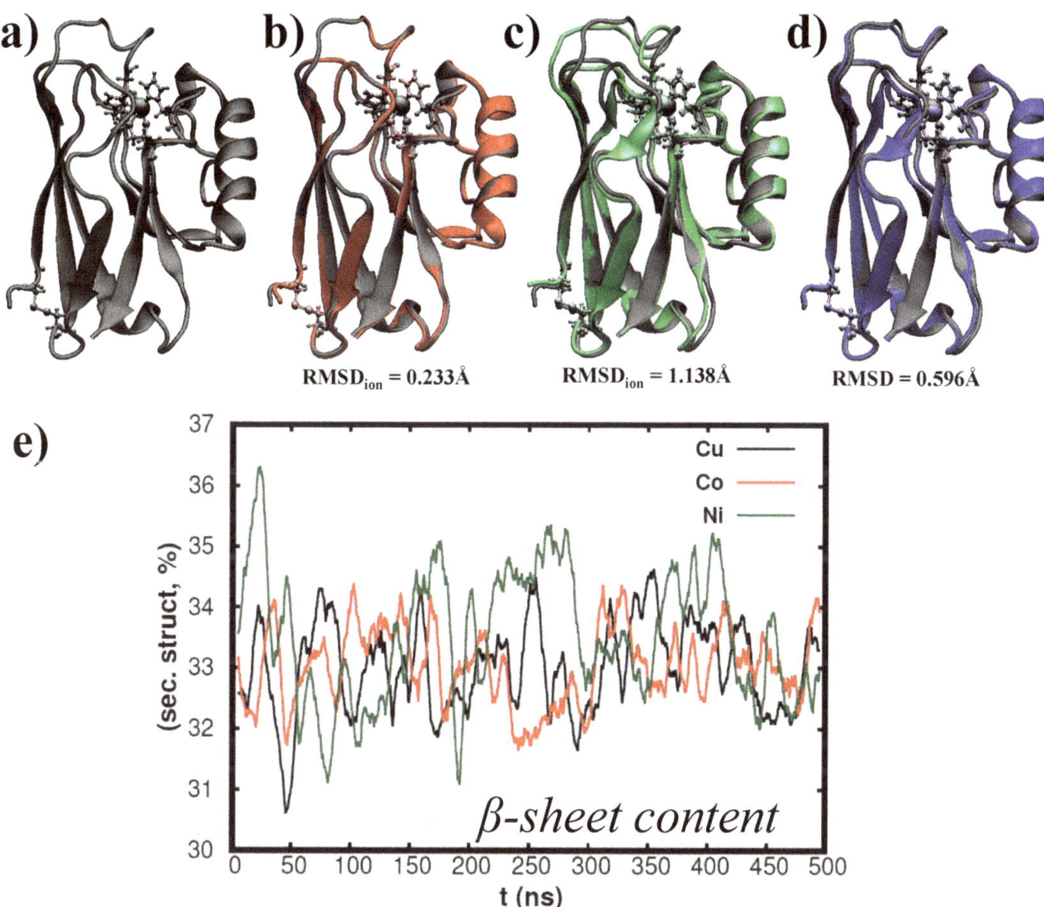

Figure 1. Structural characterization of the X-azurin (X = Cu, Co and Ni) free solved in TIP3P water obtained via atomistic MD simulations. (**a–c**) Energetically equilibrated averaged configurations obtained over the last 200 ns of MD simulation for: (**a**) native azurin; (**b**) Co-azurin; and (**c**) Ni-azurin variants. The Zn-azurin is not included given the lack of reliable force-field parameters for this variant. Note that the Co-azurin (red) and Ni-azurin (green) configurations are aligned with the native one (black) and superposed to it. The difference is quantified by the RMSD between both averaged configurations. The protein representation used in all figures is the same as in Figure 2. A comparison between the averaged Cu-azurin obtained in this work (black) with the one obtained in previous MD simulations where a different force-field for the cooper coordination sphere was used [3] is included on (**d**). (**e**) Percentage of β-sheet content for the three azurin variants considered. It is calculated using the DSSP algorithm [66] included on the CPPTRAJ ambertools [67]. Note that the β-sheet content of the azurin averaged structures is also included as the protein is represented with its secondary structure (β-sheets represented as planar arrows).

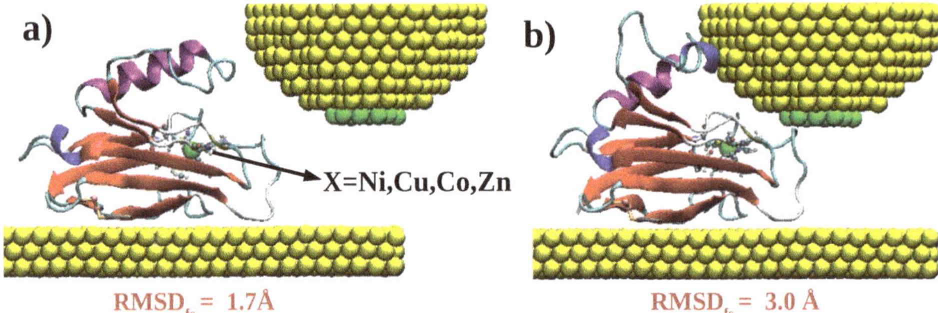

Figure 2. Initial (**a**) and final (**b**) geometries obtained during the simulations of the lateral STM-tip indentation carried out for the Cu-azurin. The arrow indicates the central copper ion which has been replaced with three different metals to reproduce the other junctions under study. The azurin is represented with its secondary structure: β-sheet (red arrows), α-helix (purple helix), 310-helix (dark-blue helix), turns (cyan) and random-coils (white). The ion is shown using its van der Waals representation in an opaque green color, and its coordination residues are represented with a ball-stick model. The disulfide bridge and the main chain of the two cysteines which formed it are colored in light orange. The sulfur atoms of these two cysteines are highlight in pink. The variation of the protein structure on the junction is quantified by computing its root-mean-square-deviation from the Cu-azurin free solved in water structure (RMSD$_{fs}$) [3].

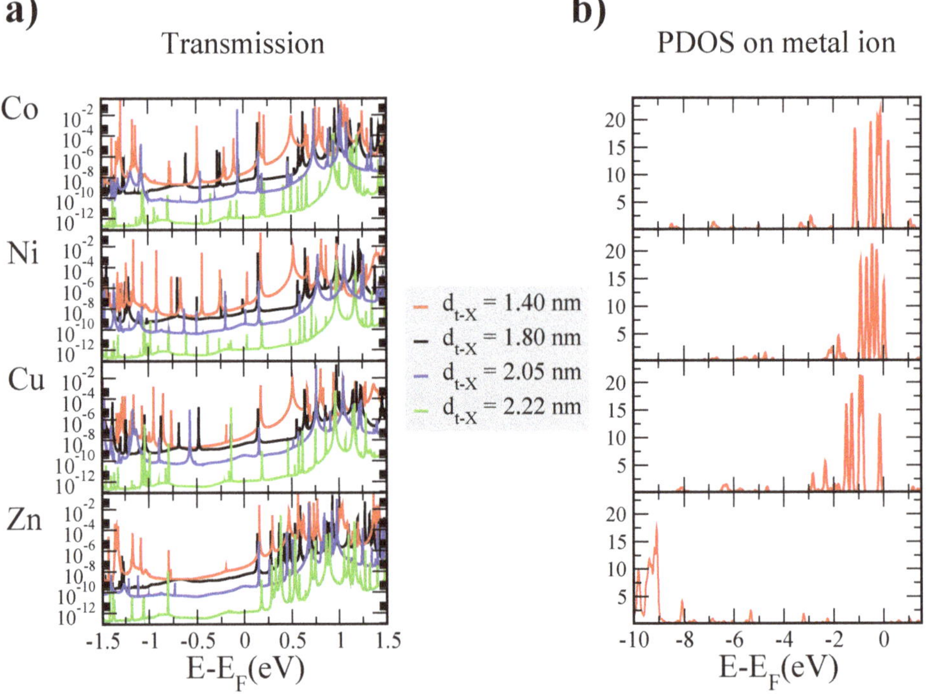

Figure 3. (**a**) Transmission as a function of energy for the four metalloproteins studied at the four different tip-protein distance values considered; and (**b**) projected density of states on the metal ion in each protein for the shortest tip-ion distance (1.40 nm). For each row, the corresponding metal ion is indicated on the left-hand side.

In the right column of the same figure, we report the projected density of states (PDOS) on the metal ion for the shortest tip-protein distance. The curve for Zn clearly differs from the other three in that the *d*-level peaks lie at lower energies with respect to the Fermi level as compared to the other metal ions (this, in turn, derives from the closed-shell electronic Zn configuration). This difference is reflected in the corresponding transmission curve, in which no peaks appear in a broad range immediately below the Fermi level as opposed to the other three proteins. Indeed, the lower density currents measured for the Zn-protein junctions and reported by Amdursky et al. [15] were ascribed to the different energy alignment of the metal-derived levels. A common feature to the transmission curves of all proteins is the energetic position of the lowest occupied orbital (localized on HIS35), which shows minimal changes from one curve to another. This is not surprising because, in our simulations, the position of the amino acids (including HIS35) within the protein is the same for all cases. Consequently, they contribute to the transmission with similar features. Note that, although the lack of re-optimization might in principle lead to some artifacts, the MD simulations discussed above revealed minimal changes upon ion replacement. This suggests that they would result into energy shifts of individual orbitals the effect of which would most likely be hidden by the rest of the contributions of the whole electronic structure of the protein (given the extreme sharpness of all peaks in the transmission curves of Figure 3, it is hard to believe that any shift of one of these peaks would affect the transmission at the Fermi level). The same applies to possible shifts of individual peaks arising from the spin polarization of the metallic ion which could not be described in our closed-shell calculations.

Figure 4 (left) shows a comparison for the transmission curves obtained for the four proteins at the shortest tip-ion distance in a narrow energy range around the Fermi level. The transmission curve for the Apo (the protein deprived of the metal ion) is also shown in the same graph. It mainly differs from the other curves in the obvious absence of the peaks which derive, in the case of the other four proteins, from the d states of the metal ions. Quite strikingly, the transmission at the Fermi level for the metal-substituted azurins and the Apo is almost identical. This is due to the extreme sharpness of the HOMO-related resonances deriving from the metal ions. Similar conclusions can be drawn for the other three tip-protein distance values (see Figure 4 (right) for a comparison of the transmission values at the Fermi level at all distances for all proteins). In the case of Co, Ni and Cu, these results suggest that, despite these metal ions contributing to levels close to the Fermi level, their role in a fully-coherent electron transport is certainly important but not so dominant as one might naively expect. A large part of the contribution seems instead to be given by the rest of the protein, which provides the same background transmission for the Apo and for the other proteins analyzed in this study. For the short-distance structure corresponding to the curves shown in Figure 4, the main role seems to be played by the LUMO (this is less visible in the case of the curves of Figure 3 corresponding to larger distances, which seem to rather indicate an equal contribution between HOMO and LUMO). In any case, one has to bear in mind that the occupied energy range immediately below the Fermi level includes the contribution of several residues (mostly glutammic and aspartic acids: for a detailed analysis in the case of the Cu-azurin, we refer the reader to the work of Romero-Muñiz et al. [34]). In the case of the Zn-azurin and for the Apo, part of the contribution to the transmission at the Fermi level seems indeed to be provided by these residues as well. As an aside, it is worth mentioning that recent experiments on protoporphyrins have revealed that the metal redox center has no role in the electron transfer, which was instead found to be mediated solely by the conjugated backbone of the molecule [68]. Our results are based on DFT, which is known to be affected by uncertainties concerning the size of the HOMO–LUMO gap because of self-interaction errors in the standard exchange-correlation functionals and image charge effects [69–74]. Nevertheless, it is hard to foresee that the main conclusions would change upon use of a more sophisticated exchange-correlation functional. Although in this study we focused on sideways approach of the tip, other mechanisms would be possible such as indentation from the top. Geometries obtained via this kind of simulation, however, would probably lead to similar conclusions as those drawn for the geometries discussed above. Such a prediction is based on the fact that, for the Cu-azurin for instance, the main features of the transmission curves (concerning coupling and energetic alignment) were found to be quite similar [34].

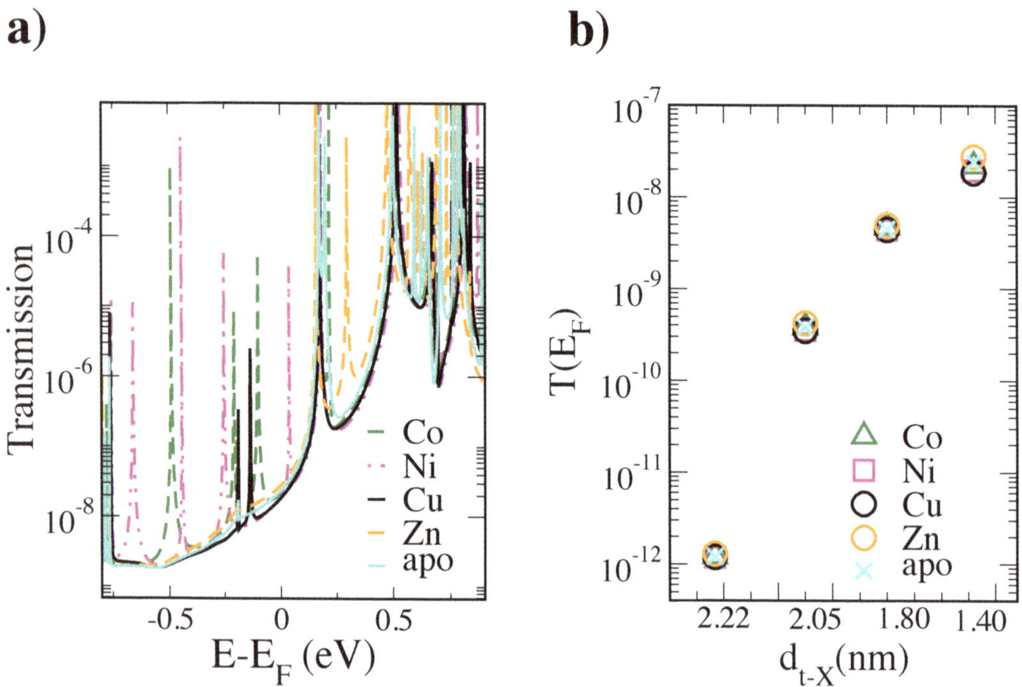

Figure 4. (**a**) Transmission as a function of energy for all five proteins studied at the shortest tip-ion distance (1.40 nm); and (**b**) transmission at the Fermi level for all five proteins at all four tip-protein distance values considered.

We now turn to comparing our results with the experimental ones reported in [15]. There, the current density was found to follow the trend Cu > Ni > Co > Zn. This is at odds with our results which, as explained above, did not show significant changes for the transmission at the Fermi level. However, our findings, based on single-protein junctions, cannot be directly compared with those experiments, which were carried out on self-assembled monolayers; the latter are affected by dipole–dipole interaction and other effects [75] which are not present in our systems. Moreover, the electrodes which were there used (a highly-doped Si substrate and, in certain cases, a top electrode made of Hg) are very different from the gold electrodes employed in our simulations. Amdursky et al. [15] also reported a decrease of two orders of magnitude for the current through the Apo with respect to the Cu-azurin for a large range of temperatures. However, we also note that, in [11], the Apo was actually found to show similar conductance to that of Cu-azurin, although this was attributed to different tunneling distances. In any case, the discrepancies between our results and those reported in [15] suggest that probably the electron-transport mechanism taking place in these systems is not as completely coherent as was claimed in multiple works [8,11,23,26,32,36–39]. It also highlights the fact that probably a proper description of the transport process needs other effects to be taken into account such as the time dependence of the fluctuations of the surrounding ligands. Similar conclusions were actually drawn by Romero-Muñiz et al. [34] concerning the Cu-azurin on the basis of the very low transmission values obtained for a broad range of geometries. We believe that these latest results of ours further confirm the need of further investigation regarding the nature the charge transfer in these kinds of junctions.

4. Conclusions

In summary, we studied metal–protein–metal junctions based on modified blue-copper azurins in which the central copper ion was replaced by Zn, Ni or Co, while leaving the rest of the structure unmodified. Comparison with the Apo protein was also performed. The striking similarity of the transmission values at the Fermi level among the five proteins suggest that, within a fully-coherent-transport picture, the dominant role would not actually be played by the metal ions, opposite to what common wisdom would suggest. The discrepancy with recent experimental observations suggest that a different kind of electron-transport mechanism is probably involved, although simulations employing more similar structures to those formed in the experiments of Amdursky et al. [15] should be carried out in order to draw definitive conclusions. Molecular dynamics simulations performed on the Cu-, Ni- and Co-azurin indicate that the ion substitutions do not alter the secondary structure significantly. The structural variation induced is much smaller than that arising from contact with the top electrode. Understanding whether such a difference can result in major changes in the conductance will be pursued in future studies.

Supplementary Materials: The following are available online at https://www.mdpi.com/article/10.3390/app11093732/s1, Figures S1–S3: partial charges; Figure S4: bond distances and force constants; Figure S5: transmission curves for the Apo azurin.

Author Contributions: Conceptualization, C.R.-M. and L.A.Z.; Funding acquisition, C.R.-M., J.C.C., R.P. and L.A.Z.; Investigation, C.R.-M., M.O. and L.A.Z.; Resources, R.P. and L.A.Z.; Writing—original draft, C.R.-M. and L.A.Z.; and Writing—review and editing, C.R.-M., M.O., J.G.V., J.C.C., R.P. and L.A.Z. All authors have read and agreed to the published version of the manuscript.

Funding: C.R.-M. acknowledges funding from the Spanish MICINN via the Juan de la Cierva-Formación program (Ref. FJC2018-036832). J.G.V. acknowledges funding from a Marie Sklodowska-Curie Fellowship within the Horizons 2020 framework (grant number DLV-795286), and the Swiss National Science Foundation (grant number CRSK-2_190731/1). R.P. and J.C.C. acknowledge funding from the Spanish MINECO (Contract Nos. MAT2017-83273-R, FIS2017-84057-P and "María de Maeztu" Programme for Units of Excellence in R&D (grant No. CEX2018-000805-M)). L.A.Z. thanks financial support from the University of Seville through the VI PPIT-US program.

Institutional Review Board Statement: Not applicable

Informed Consent Statement: Not applicable

Data Availability Statement: All the data and additional information supporting the findings of this study are available from the corresponding authors upon reasonable request.

Acknowledgments: The authors thankfully acknowledge the computer resources, technical expertise and assistance provided by the Red Espa nola de Supercomputación (RES) at the Minotauro and Marenostrum (BSC, Barcelona) and Calendula (SCAYLE, Castilla y León) supercomputers and by Centro Informático Científico de Andalucía (CICA).

Conflicts of Interest: The authors declare no conflict of interest.

Abbreviations

The following abbreviations are used in this manuscript:

STM	Scanning Tunneling Microscopy
MD	Molecular Dynamics
DFT	Density Functional Theory
RMSD	Root-Mean-Square-Deviation
LUMO	Lowest Unoccupied Molecular Orbital
HOMO	Highest Occupied Molecular Orbital

References

1. Panda, S.S.; Katz, H.E.; Tovar, J.D. Solid-state Electrical Applications of Protein and Peptide Based Nanomaterials. *Chem. Soc. Rev.* **2018**, *47*, 3640–3658. [CrossRef]
2. Chi, Q.; Farver, O.; Ulstrup, J. Long-Range Protein Electron Transfer Observed at the Single-Molecule Level: In Situ Mapping of Redox-Gated Tunneling Resonance. *Proc. Natl. Acad. Sci. USA* **2005**, *102*, 16203–16208. [CrossRef] [PubMed]
3. Ortega, M.; Vilhena, J.G.; Zotti, L.A.; Díez-Pérez, I.; Cuevas, J.C.; Pérez, R. Tuning Structure and Dynamics of Blue Copper Azurin Junctions via Single Amino-Acid Mutations. *Biomolecules* **2019**, *9*, 611. [CrossRef]
4. Bostick, C.D.; Mukhopadhyay, S.; Pecht, I.; Sheves, M.; Cahen, D.; Lederman, D. Protein Bioelectronics: A Review of what We Do and Do not Know. *Rep. Prog. Phys.* **2018**, *81*, 026601. [CrossRef] [PubMed]
5. Ing, N.L.; El-Naggar, M.Y.; Hochbaum, A.I. Going the distance: Long-range conductivity in protein and peptide bioelectronic materials. *J. Phys. Chem. B* **2018**, *122*, 10403–10423. [CrossRef]
6. Alessandrini, A.; Facci, P. Electron Transfer in Nanobiodevices. *Eur. Polym. J.* **2016**, *83*, 450–466. [CrossRef]
7. Eleonora, A.; Reggiani, L.; Pousset, J. Proteotronics: Electronic devices based on proteins. In *Sensors*; Springer: New York, NY, USA, 2015; pp. 3–7.
8. Fereiro, J.A.; Kayser, B.; Romero-Muñiz, C.; Vilan, A.; Dolgikh, D.A.; Chertkova, R.V.; Cuevas, J.C.; Zotti, L.A.; Pecht, I.; Sheves, M.; et al. A Solid-State Protein Junction Serves as a Bias-Induced Current Switch. *Angew. Chem. Int. Ed.* **2019**, *58*, 11852–11859. [CrossRef]
9. Futera, Z.; Ide, I.; Kayser, B.; Garg, K.; Jiang, X.; van Wonderen, J.H.; Butt, J.N.; Ishii, H.; Pecht, I.; Sheves, M.; et al. Coherent Electron Transport across a 3 nm Bioelectronic Junction Made of Multi-Heme Proteins. *J. Phys. Chem. Lett.* **2020**, *11*, 9766–9774. [CrossRef]
10. Ruiz, M.P.; Aragonès, A.C.; Camarero, N.; Vilhena, J.G.; Ortega, M.; Zotti, L.A.; Pérez, R.; Cuevas, J.C.; Gorostiza, P.; Díez-Pérez, I. Bioengineering a Single-Protein Junction. *J. Am. Chem. Soc.* **2017**, *139*, 15337–15346. [CrossRef]
11. Fereiro, J.A.; Yu, X.; Pecht, I.; Sheves, M.; Cuevas, J.C.; Cahen, D. Tunneling Explains Efficient Electron Transport via Protein Junctions. *Proc. Natl. Acad. Sci. USA* **2018**, *115*, E4577. [CrossRef]
12. Zotti, L.A.; Bednarz, B.; Hurtado-Gallego, J.; Cabosart, D.; Rubio-Bollinger, G.; Agrait, N.; van der Zant, H.S.J. Can One Define the Conductance of Amino Acids? *Biomolecules* **2019**, *9*, 580. [CrossRef]
13. Zotti, L.A.; Cuevas, J.C. Electron Transport Through Homopeptides: Are They Really Good Conductors? *ACS Omega* **2018**, *3*, 3778–3785. [CrossRef]
14. Schosser, W.M.; Zotti, L.A.; Cuevas, J.C.; Pauly, F. Doping hepta-alanine with tryptophan: A theoretical study of its effect on the electrical conductance of peptide-based single-molecule junctions. *J. Chem. Phys.* **2019**, *150*, 174705. [CrossRef]
15. Amdursky, N.; Sepunaru, L.; Raichlin, S.; Pecht, I.; Sheves, M.; Cahen, D. Electron Transfer Proteins as Electronic Conductors: Significance of the Metal and Its Binding Site in the Blue Cu Protein, Azurin. *Adv. Sci.* **2015**, *2*, 1400026. [CrossRef] [PubMed]
16. Liu, Z.F.; Wei, S.; Yoon, H.; Adak, O.; Ponce, I.; Jiang, Y.; Jang, W.D.; Campos, L.M.; Venkataraman, L.; Neaton, J.B. Control of Single-Molecule Junction Conductance of Porphyrins via a Transition-Metal Center. *Nano Lett.* **2014**, *14*, 5365–5370. [CrossRef] [PubMed]
17. Mahapatro, A.K.; Ying, J.; Ren, T.; Janes, D.B. Electronic Transport through Ruthenium-Based Redox-Active Molecules in Metal-Molecule-Metal Nanogap Junctions. *Nano Lett.* **2008**, *8*, 2131–2136. [CrossRef] [PubMed]
18. Zotti, L.A.; Leary, E.; Soriano, M.; Cuevas, J.C.; Palacios, J.J. A Molecular Platinum Cluster Junction: A Single-Molecule Switch. *J. Am. Chem. Soc.* **2013**, *135*, 2052–2055. [CrossRef]
19. McLaughlin, M.P.; Retegan, M.; Bill, E.; Payne, T.M.; Shafaat, H.S.; Peña, S.; Sudhamsu, J.; Ensign, A.A.; Crane, B.R.; Neese, F.; et al. Azurin as a Protein Scaffold for a Low-coordinate Nonheme Iron Site with a Small-molecule Binding Pocket. *J. Am. Chem. Soc.* **2012**, *134*, 19746–19757. [CrossRef] [PubMed]
20. Rajapandian, V.; Hakkim, V.; Subramanian, V. Molecular Dynamics Studies on Native, Loop-Contracted, and Metal Ion-Substituted Azurins. *J. Phys. Chem. B* **2010**, *114*, 8474–8486. [CrossRef]
21. Solomon, E.I.; Szilagyi, R.K.; DeBeer George, S.; Basumallick, L. Electronic Structures of Metal Sites in Proteins and Models: Contributions to Function in Blue Copper Proteins. *Chem. Rev.* **2004**, *104*, 419. [CrossRef]
22. Ron, I.; Sepunaru, L.; Itzhakov, S.; Belenkova, T.; Friedman, N.; Pecht, I.; Sheves, M.; Cahen, D. Proteins as Electronic Materials: Electron Transport through Solid-State Protein Monolayer Junctions. *J. Am. Chem. Soc.* **2010**, *132*, 4131–4140. [CrossRef] [PubMed]
23. Sepunaru, L.; Pecht, I.; Sheves, M.; Cahen, D. Solid-State Electron Transport across Azurin: From a Temperature-Independent to a Temperature-Activated Mechanism. *J. Am. Chem. Soc.* **2011**, *133*, 2421–2423. [CrossRef]
24. Yu, X.; Lovrincic, R.; Sepunaru, L.; Li, W.; Vilan, A.; Pecht, I.; Sheves, M.; Cahen, D. Insights into Solid-State Electron Transport through Proteins from Inelastic Tunneling Spectroscopy: The Case of Azurin. *ACS Nano* **2015**, *9*, 9955–9963. [CrossRef]
25. Fereiro, J.A.; Porat, G.; Bendikov, T.; Pecht, I.; Sheves, M.; Cahen, D. Protein Electronics: Chemical Modulation of Contacts Control Energy Level Alignment in Gold-Azurin-Gold Junctions. *J. Am. Chem. Soc.* **2018**, *140*, 13317–13326. [CrossRef]
26. Kumar, K.S.; Pasula, R.R.; Lim, S.; Nijhuis, C.A. Long-Range Tunneling Processes across Ferritin-Based Junctions. *Adv. Mater.* **2016**, *28*, 1824–1830. [CrossRef] [PubMed]
27. Mukhopadhyay, S.; Karuppannan, S.K.; Guo, C.; Fereiro, J.A.; Bergren, A.; Mukundan, V.; Qiu, X.; Casta neda Ocampo, O.E.; Chen, X.; Chiechi, R.C.; et al. Solid-State Protein Junctions: Cross-Laboratory Study Shows Preservation of Mechanism at Varying Electronic Coupling. *iScience* **2020**, *23*, 101099. [CrossRef]

28. Artés, J.M.; Díez-Pérez, I.; Gorostiza, P. Transistor-like Behavior of Single Metalloprotein Junctions. *Nano Lett.* **2012**, *12*, 2679–2684. [CrossRef]
29. Artés, J.M.; López-Martínez, M.; Díez-Pérez, I.; Sanz, F.; Gorostiza, P. Conductance Switching in Single Wired Redox Proteins. *Small* **2014**, *10*, 2537–2541. [CrossRef] [PubMed]
30. Li, W.; Sepunaru, L.; Amdursky, N.; Cohen, S.R.; Pecht, I.; Sheves, M.; Cahen, D. Temperature and Force Dependence of Nanoscale Electron Transport via the Cu Protein Azurin. *ACS Nano* **2012**, *6*, 10816–10824. [CrossRef]
31. Baldacchini, C.; Bizzarri, A.R.; Cannistraro, S. Electron Transfer, Conduction and Biorecognition Properties of the Redox Metalloprotein Azurin Assembled onto Inorganic Substrates. *Eur. Polym. J.* **2016**, *83*, 407–427. [CrossRef]
32. Kayser, B.; Fereiro, J.A.; Bhattacharyya, R.; Cohen, S.R.; Vilan, A.; Pecht, I.; Sheves, M.; Cahen, D. Solid-State Electron Transport via the Protein Azurin is Temperature-Independent Down to 4 K. *J. Phys. Chem. Lett.* **2020**, *11*, 144–151. [CrossRef] [PubMed]
33. Zhao, J.; Davis, J.J.; Sansom, M.S.P.; Hung, A. Exploring the Electronic and Mechanical Properties of Protein Using Conducting Atomic Force Microscopy. *J. Am. Chem. Soc.* **2004**, *126*, 5601–5609. [CrossRef]
34. Romero-Muñiz, C.; Ortega, M.; Vilhena, J.G.; Díez-Pérez, I.; Pérez, R.; Cuevas, J.C.; Zotti, L.A. Can Electron Transport through a Blue-Copper Azurin Be Coherent? An Ab Initio Study. *J. Phys. Chem. C* **2021**, *125*, 1693–1702. [CrossRef]
35. Valianti, S.; Cuevas, J.C.; Skourtis, S.S. Charge-Transport Mechanisms in Azurin-Based Monolayer Junctions. *J. Phys. Chem. C* **2019**, *123*, 5907–5922. [CrossRef]
36. Mukhopadhyay, S.; Dutta, S.; Pecht, I.; Sheves, M.; Cahen, D. Conjugated Cofactor Enables Efficient Temperature-Independent Electronic Transport across 6 nm Long Halorhodopsin. *J. Am. Chem. Soc.* **2015**, *137*, 11226–11229. [CrossRef] [PubMed]
37. Casta neda Ocampo, O.E.; Gordiichuk, P.; Catarci, S.; Gautier, D.A.; Herrmann, A.; Chiechi, R.C. Mechanism of Orientation-Dependent Asymmetric Charge Transport in Tunneling Junctions Comprising Photosystem I. *J. Am. Chem. Soc.* **2015**, *137*, 8419–8427. [CrossRef] [PubMed]
38. Garg, K.; Ghosh, M.; Eliash, T.; van Wonderen, J.H.; Butt, J.N.; Shi, L.; Jiang, X.; Zdenek, F.; Blumberger, J.; Pecht, I.; et al. Direct Evidence for Heme-Assisted Solid-State Electronic Conduction in Multi-Heme c-Type Cytochromes. *Chem. Sci.* **2018**, *9*, 7304–7310. [CrossRef]
39. Garg, K.; Raichlin, S.; Bendikov, T.; Pecht, I.; Sheves, M.; Cahen, D. Interface Electrostatics Dictates the Electron Transport via Bioelectronic Junctions. *ACS Appl. Mater. Interfaces* **2018**, *10*, 41599–41607. [CrossRef]
40. Blaszak, J.A.; Ulrich, E.L.; Markley, J.L.; McMillin, D.R. High-resolution proton nuclear magnetic resonance studies of the nickel(II) derivative of azurin. *Biochemistry* **1982**, *21*, 6253–6258. [CrossRef]
41. Czernuszewicz, R.S.; Fraczkiewicz, G.; Zareba, A.A. A Detailed Resonance Raman Spectrum of Nickel(II)-Substituted Pseudomonas aeruginosa Azurin. *Inorg. Chem.* **2005**, *44*, 5745–5752. [CrossRef]
42. Jiménez, H.R.; Salgado, J.; Moratal, J.M.; Morgenstern-Badarau, I. EPR and Magnetic Susceptibility Studies of Cobalt(II)- and Nickel(II)-Substituted Azurins from Pseudomonas aeruginosa. Electronic Structure of the Active Sites. *Inorg. Chem.* **1996**, *35*, 2737–2741. [CrossRef]
43. Nar, H.; Huber, R.; Messerschmidt, A.; Filippou, A.C.; Barth, M.; Jaquind, M.; van de Kamp, M.; Canters, G.W. Characterization and crystal structure of zinc azurin, a by-product of heterologous expression in Escherichia coli of Pseudomonas aeruginosa copper azurin. *Eur. J. Biochem.* **1992**, *205*, 1123–1129. [CrossRef] [PubMed]
44. Bonander, N.; Vänngård, T.; Tsai, L.C.; Langer, V.; Nar, H.; Sjölin, L. The metal site of Pseudomonas aeruginosa azurin, revealed by a crystal structure determination of the co(II) derivative and co-EPR spectroscopy. *Proteins Struct. Funct. Bioinf.* **1997**, *27*, 385. [CrossRef]
45. Moratal, J.M.; Romero, A.; Salgado, J.; Perales-Alarcón, A.; Jiménez, H.R. The Crystal Structure of Nickel(II)-Azurin. *Eur. J. Biochem.* **1995**, *228*, 653. [CrossRef]
46. Berman, H.M.; Westbrook, J.; Feng, Z.; Gilliland, G.; Bhat, T.N.; Weissig, H.; Shindyalov, I.N.; Bourne, P.E. The Protein Data Bank. *Nucleic Acids Res.* **2000**, *28*, 235–242. [CrossRef]
47. Nar, H.; Messerschmidt, A.; Huber, R.; van de Kamp, M.; Canters, G.W. Crystal Structure Analysis of Oxidized Pseudomonas Aeruginosa Azurin at pH 5.5 and pH 9.0: A pH-Induced Conformational Transition Involves a Peptide Bond Flip. *J. Mol. Biol.* **1991**, *221*, 765–772. [CrossRef]
48. Gordon, J.C.; Myers, J.B.; Folta, T.; Shoja, V.; Heath, L.S.; Onufriev, A. H++: A Server for Estimating pKas and Adding Missing Hydrogens to Macromolecules. *Nucleic Acids Res.* **2005**, *33*, W368–W371. [CrossRef]
49. Maier, J.A.; Martinez, C.; Kasavajhala, K.; Wickstrom, L.; Hauser, K.E.; Simmerling, C. ff14SB: Improving the Accuracy of Protein Side Chain and Backbone Parameters from ff99SB. *J. Chem. Theory Comput.* **2015**, *11*, 3696–3713. [CrossRef] [PubMed]
50. Jorgensen, W.L.; Chandrasekhar, J.; Madura, J.D.; Impey, R.W.; Klein, M.L. Comparison of Simple Potential Functions for Simulating Liquid Water. *J. Chem. Phys.* **1983**, *79*, 926–935. [CrossRef]
51. Case, D.A.; Darden, T.A.; Simmerling, C.; Wang, J.; Duke, R.; Luo, R.; Walker, R.; Zhang, W.; Merz, K.; Roberts, B.; et al. *AMBER 14*; University of California: San Francisco, CA, USA, 2014. Available online: http://ambermd.org/ (accessed on June 15, 2018).
52. Salomon-Ferrer, R.; Götz, A.W.; Poole, D.; Le Grand, S.; Walker, R.C. Routine Microsecond Molecular Dynamics Simulations with AMBER on GPUs. 2. Explicit Solvent Particle Mesh Ewald. *J. Chem. Theory Comput.* **2013**, *9*, 3878–3888. [CrossRef]
53. Götz, A.W.; Williamson, M.J.; Xu, D.; Poole, D.; Le Grand, S.; Walker, R.C. Routine Microsecond Molecular Dynamics Simulations with AMBER on GPUs. 1. Generalized Born. *J. Chem. Theory Comput.* **2012**, *8*, 1542–1555. [CrossRef]

54. Grand, S.L.; Götz, A.W.; Walker, R.C. SPFP: Speed without Compromise: A Mixed Precision Model for GPU Accelerated Molecular Dynamics Simulations. *Comput. Phys. Commun.* **2013**, *184*, 374–380. [CrossRef]
55. Loncharich, R.J.; Brooks, B.R.; Pastor, R.W. Langevin dynamics of peptides: The frictional dependence of isomerization rates of N-acetylalanyl-N-methylamide. *Biopolymers* **1992**, *32*, 523. [CrossRef]
56. Berendsen, H.J.C.; Postma, J.P.M.; van Gunsteren, W.F.; DiNola, A.; Haak, J.R. Molecular dynamics with coupling to an external bath. *J. Chem. Phys.* **1984**, *81*, 3684. [CrossRef]
57. Miyamoto, S.; Kollman, P.A. Settle: An analytical version of the SHAKE and RATTLE algorithm for rigid water models. *J. Comput. Chem.* **1992**, *13*, 952. [CrossRef]
58. Ozaki, T. Variationally Optimized Atomic Orbitals for Large-Scale Electronic Structures. *Phys. Rev. B* **2003**, *67*, 155108. [CrossRef]
59. Ozaki, T.; Kino, H. Numerical Atomic Basis Orbitals from H to Kr. *Phys. Rev. B* **2004**, *69*, 195113. [CrossRef]
60. Romero-Mu niz, C.; Ortega, M.; Vilhena, J.G.; Díez-Pérez, I.; Cuevas, J.C.; Pérez, R.; Zotti, L.A. Ab Initio Electronic Structure Calculations of Entire Blue Copper Azurins. *Phys. Chem. Chem. Phys.* **2018**, *20*, 30392–30402. [CrossRef] [PubMed]
61. Romero-Mu niz, C.; Ortega, M.; Vilhena, J.G.; Diéz-Pérez, I.; Cuevas, J.C.; Pérez, R.; Zotti, L.A. Mechanical Deformation and Electronic Structure of a Blue Copper Azurin in a Solid-State Junction. *Biomolecules* **2019**, *9*, 506. [CrossRef]
62. Perdew, J.P.; Burke, K.; Ernzerhof, M. Generalized Gradient Approximation Made Simple. *Phys. Rev. Lett.* **1996**, *77*, 3865. [CrossRef]
63. Morrison, I.; Bylander, D.M.; Kleinman, L. Nonlocal Hermitian Norm-Conserving Vanderbilt Pseudopotential. *Phys. Rev. B* **1993**, *47*, 6728. [CrossRef]
64. Kresse, G.; Furthmüller, J. Efficient Iterative Schemes for *Ab Initio* Total-Energy Calculations Using a Plane-Wave Basis Set. *Phys. Rev. B* **1996**, *54*, 11169–11186. [CrossRef]
65. Kerker, G.P. Efficient Iteration Scheme for Self-Consistent Pseudopotential Calculations. *Phys. Rev. B* **1981**, *23*, 3082–3084. [CrossRef]
66. Kabsch, W.; Sander, C. Dictionary of protein secondary structure: Pattern recognition of hydrogen-bonded and geometrical features. *Biopolymers* **1983**, *22*, 2577–2637. [CrossRef] [PubMed]
67. Roe, D.R.; Cheatham, T.E. PTRAJ and CPPTRAJ: Software for Processing and Analysis of Molecular Dynamics Trajectory Data. *J. Chem. Theory Comput.* **2013**, *9*, 3084–3095. [CrossRef]
68. Agam, Y.; Nandi, R.; Kaushansky, A.; Peskin, U.; Amdursky, N. The porphyrin ring rather than the metal ion dictates long-range electron transport across proteins suggesting coherence-assisted mechanism. *Proc. Natl. Acad. Sci. USA* **2020**, *117*, 32260–32266. [CrossRef]
69. Jakobsen, S.; Kristensen, K.; Jensen, F. Electrostatic Potential of Insulin: Exploring the Limitations of Density Functional Theory and Force Field Methods. *J. Chem. Theory Comput.* **2013**, *9*, 3978–3985. [CrossRef] [PubMed]
70. Zhao, Q.; Kulik, H.J. Where Does the Density Localize in the Solid State? divergent behavior for hybrids and DFT+ U. *J. Chem. Theory Comput.* **2018**, *14*, 670–683. [CrossRef] [PubMed]
71. Bao, J.L.; Gagliardi, L.; Truhlar, D.G. Self-Interaction Error in Density Functional Theory: An Appraisal. *J. Phys. Chem. Lett.* **2018**, *9*, 2353–2358. [CrossRef]
72. Zotti, L.A.; Bürkle, M.; Pauly, F.; Lee, W.; Kim, K.; Jeong, W.; Asai, Y.; Reddy, P.; Cuevas, J.C. Heat dissipation and its relation to thermopower in single-molecule junctions. *New J. Phys.* **2014**, *16*, 015004. [CrossRef]
73. Montes, E.; Vázquez, H. Role of the Binding Motifs in the Energy Level Alignment and Conductance of Amine-Gold Linked Molecular Junctions within DFT and DFT + Σ. *Appl. Sci.* **2021**, *11*, 802. [CrossRef]
74. Garcia-Lastra, J.M.; Rostgaard, C.; Rubio, A.; Thygesen, K.S. Polarization-induced renormalization of molecular levels at metallic and semiconducting surfaces. *Phys. Rev. B* **2009**, *80*, 245427. [CrossRef]
75. Rissner, F.; Natan, A.; Egger, D.A.; Hofmann, O.T.; Kronik, L.; Zojer, E. Dimensionality effects in the electronic structure of organic semiconductors consisting of polar repeat units. *Org. Electron.* **2012**, *13*, 3165–3176. [CrossRef] [PubMed]

Article

Tuning Single-Molecule Conductance by Controlled Electric Field-Induced *trans*-to-*cis* Isomerisation

C.S. Quintans [1], Denis Andrienko [2], Katrin F. Domke [2], Daniel Aravena [3,*], Sangho Koo [4], Ismael Díez-Pérez [5,*] and Albert C. Aragonès [5,*,†]

[1] Department of Chemistry, Federal University of São Carlos, São Carlos 13565-905, Brazil; ciroquintans+papers@gmail.com

[2] Max Planck Institute for Polymer Research, Ackermannweg 10, 55128 Mainz, Germany; denis.andrienko@mpip-mainz.mpg.de (D.A.); domke@mpip-mainz.mpg.de (K.F.D.)

[3] Departamento de Química de los Materiales, Facultad de Química y Biología, Universidad de Santiago de Chile, Casilla 40, Correo 33, Santiago 9170022, Chile

[4] Department of Chemistry, Myongji University, Myongji-Ro 116, Cheoin-Gu, Yongin 17058, Gyeonggi-Do, Korea; sangkoo@mju.ac.kr

[5] Department of Chemistry, Faculty of Natural & Mathematical Sciences, King's College London Britannia House, 7 Trinity Street, London SE1 1DB, UK

* Correspondence: daniel.aravena.p@usach.cl (D.A.); ismael.diez_perez@kcl.ac.uk (I.D.-P.); albert.cortijos@mpip-mainz.mpg.de (A.C.A.)

† Current address: Max Planck Institute for Polymer Research, Ackermannweg 10, 55128 Mainz, Germany.

Abstract: External electric fields (EEFs) have proven to be very efficient in catalysing chemical reactions, even those inaccessible via wet-chemical synthesis. At the single-molecule level, oriented EEFs have been successfully used to promote in situ single-molecule reactions in the absence of chemical catalysts. Here, we elucidate the effect of an EEFs on the structure and conductance of a molecular junction. Employing scanning tunnelling microscopy break junction (STM-BJ) experiments, we form and electrically characterize single-molecule junctions of two tetramethyl carotene isomers. Two discrete conductance signatures show up more prominently at low and high applied voltages which are univocally ascribed to the *trans* and *cis* isomers of the carotenoid, respectively. The difference in conductance between both cis-/trans- isomers is in concordance with previous predictions considering π-quantum interference due to the presence of a single *gauche* defect in the trans isomer. Electronic structure calculations suggest that the electric field polarizes the molecule and mixes the excited states. The mixed states have a (spectroscopically) allowed transition and, therefore, can both promote the cis-isomerization of the molecule and participate in electron transport. Our work opens new routes for the in situ control of isomerisation reactions in single-molecule contacts.

Keywords: molecular electronics; single-molecule junctions; STM break-junction; in-situ isomerisation; carotenoids

Citation: Quintans, C.S.; Andrienko, D.; Domke, K.F.; Aravena, D.; Koo, S.; Díez-Pérez, I.; Aragonès, A.C. Tuning Single-Molecule Conductance by Controlled Electric Field-Induced *trans*-to-*cis* Isomerisation. *Appl. Sci.* **2021**, *11*, 3317. https://doi.org/10.3390/app11083317

Academic Editor: Linda Angela Zotti

Received: 13 March 2021
Accepted: 5 April 2021
Published: 7 April 2021

1. Introduction

The development of novel, more efficient ways to control molecular reactions has been a restless quest for synthetic chemists. Many different triggers, such as light, heat or external electric fields (EEF) are being used to promote chemical reactions [1,2]. EEF have been shown to be able to stabilize conventionally non-favourable electronic structures [3] and thus to enable new transition states [4,5] in a theoretically predictable manner. These field-induced chemical reactivity experiments, which are well reported in single-molecule devices [3,5], represent an exciting alternative to traditional bulk chemical catalyst approaches since EEF provide a cost-, material-efficient methodology to precisely control molecular reactions in a cleaner, more sustainable way.

In the molecular electronics (ME) field, one of the most appealing single-molecule reactions is the isomerisation reaction [6–9], because small changes in the molecular structure give rise to a significant change in the conductance of the molecular wire [8,9]. The molecular configuration constitutes a defining parameter for the intrinsic conducting properties of a molecule. Moreover, measuring single-molecule conductance can be used to follow structural changes in a molecular contact [10–13], opening the way to real time detection of electrical currents associated with in situ stimuli-induced molecular structural changes [5,14,15]. Charge transport across molecular systems as a function of molecule configuration has been widely studied theoretically [16–18] and experimentally [8,9,11,19–22], including *cis-/trans-* isomerization. Different isomers exhibit characteristic conductance values that can be attributed to various effects, such as the potential energy barrier variation due to structural change [8,17,19], the different contact geometry [8,9,16,21], or the different energies of the frontier molecular orbitals with respect to the electrode Fermi levels for each isomer [17,18,22]. Behind unravelling the particular phenomenology for each system, the detection of electrical currents across molecules in metal | molecule | metal junctions [13,23,24] have been essential to electrically characterize structural changes of individual molecular systems [25].

Here, we report on single-molecule scanning tunnelling microscopy break junction (STM-BJ) conductance experiments of all-*E*-carotene with terminal 1,5-dimethyl-3-methylthiophenyl groups (*trans*-TMC, Figure 1a) and 9-*Z*-carotene with terminal 3-methylthiophenyl groups (*cis*-TMC, Figure 1b) carotenoid derivatives. Both, *trans*-TMC and *cis*-TMC-syn molecules are chemically stable at room temperature [22,26]. Carotenoids contain a delocalized π-conjugated polyene backbone chain with planar configuration maximizing π-orbital overlap resulting in an efficient electron pathway [27–29]. They serve in several natural photosynthetic systems as energy/electron transfer mediator [30,31]. Our main findings indicate: (1) applying an STM bias voltage to the high-conductance (HC) *trans*-TMC-based molecular junction promotes the in-situ switching to a low-conductance (LC) *cis*-TMC-based molecular junction whose chemical compound cannot be synthesized wet chemically due to steric constraints. (2) The ratio of LC/HC signatures scales with the V_{bias} (i.e., EEF) magnitudes. At low biases, the weaker EEF do not promote the isomerisation and thus *trans-TMC* is the only detectable isomer. In contrast, at medium and high biases, the EEF strength is enough to promote the isomerisation and the *cis-TMC* becomes electrically detectable. (3) According to our density functional theory (DFT) calculations the isomerisation is facilitated via an EEF-induced excited state in the *trans*-TMC. (4) The difference in junction's lifetime between *cis*-TMC and *trans*-TMC isomers is attributed to a relatively high stability and more constrained configuration of the former in the molecular junction. (5) The low conductance for the cis isomer is ascribed to the presence of a single point *gauche* defect in the carotenoid alkene backbone which breaks the π-orbital pathway thus lowering conductance [32].

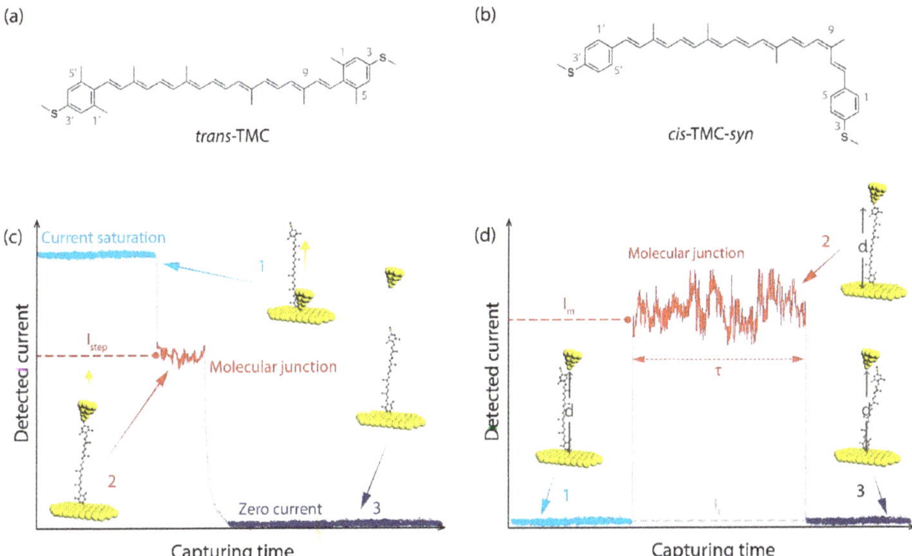

Figure 1. (**a**) *Trans*-TMC (terminal 1,5-dimethyl-3-methylthiophenyl) and (**b**) *cis*-TMC-syn synthesized molecules. (**c**) Schematic of the dynamic BJ approach: (1) current saturation; tip and substrate in contact, (2) current plateau at I_{step} due to the formation of a molecular junction during tip retraction, (3) zero current (junction breakdown) stage when the tip is pulled away from the surface and the molecule detaches from one electrode. (**d**) Schematic of the blinking approach: (1) fixed inter-electrode distance, d, at a pre-defined tunnelling current, I_t. (2) current blink at, I_m due to the spontaneous formation of a molecular junction (with a finite lifetime τ), (3) current drop to I_t upon spontaneous molecular junction breakdown.

2. Materials and Methods

2.1. Experiments

The conductance measurements were carried out with a mechanically and electronically isolated "PicoSPM II" microscope head controlled by Picoscan-2500 electronics (Agilent) using a homemade PTFE STM cell. Current signals from the STM were captured using a NI-DAQmx/BNC-2110 (National Instruments analogic-digital converter PC-interface acquisition system), analysed with LabVIEW software and plotted employing Python through Matplotlib [33]. All glassware and homemade PTFE cells were cleaned with freshly prepared piranha solution (volume ratio of 3:1 H_2SO_4:H_2O_2) before the experiments and subsequently rinsed with 18 MΩ cm^{-1} Milli-Q water (Millipore®, Burlingtone, MA, USA). A Au(111) single crystal (10 mm × 1 mm, MaTeck) of 5N purity and an orientation accuracy of <0.1° was employed as a substrate. Before each experiment, the Au(111) crystal was electro-polished to eliminate possible residual contamination, rinsed with Milli-Q water, annealed in a H_2 flame for 10 minutes and then cooled down in Ar atmosphere. The crystal was then placed in the STM cell that was filled with 80 µL of pure mesitylene (purity 99 %, ACROS Organic, Thermo Fisher Scientific). One drop of a 0.5 µM mesitylene solution containing the target carotenoid molecule was added to the STM cell before starting the single-molecule conductance experiments. The employed molecules in this work, *trans*-TMC and *cis*-TMC-syn, were synthetized according to a procedure described by Kim et al. [22] (for synthetic details see Appendix A.4).

2.2. Simulations

All calculations were performed using the ORCA 4.2.1 software package [34]. DFT and time dependent DFT TD-DFT [35] calculations with the BP86 density functional [36,37], theDef2-TZVP basis set [38,39] and the D3 corrections [40,41]. State overlaps were calculated by the WF

Overlap program [42] where excited states were constructed from all TD-DFT orbital excitations with a weight larger than 1×10^{-6}. This tight setting led to many determinants to account for the 10 first excited singlets, ranging from 4015 to 5333, depending on the calculation. Comparison between the overlap matrix before and after orthonormalisation demonstrates the adequacy of the determinant basis to account for the excited state wave functions as evidenced by the small relative angles between both matrices (0.023 and 0.004 for the *cis*-TMC and *trans*-TMC, respectively).

3. Results and Discussion

We have employed dynamic and static STM-BJ methods to characterize single-molecule TMC isomers conductance and to study the effect of the EEF generated by the STM bias voltage, on the conductance of the TMC single-molecule junction. Briefly, in the dynamic BJ approach (Figure 1c), the STM tip electrode is repeatedly moved into and out of contact with the substrate electrode at a constant piezo voltage ramp at 0.5 V/s with the STM piezo servo-feedback loop off. The junction conductivity is captured via the current versus time (or displacement) traces in the retraction stage. During the contact process (stage 1 in Figure 1c,), the two metal electrodes (STM tip and substrate) are in contact, which the corresponding current saturation due to high conductive nature of the metal/metal contact, and individual carotenoid molecules in solution spontaneously attach to them through the -SMe terminal groups that have a high affinity for Au [26]. When the tip is retracted from the substrate, the metallic contact breaks and makes it possible for a TMC molecule to bridge the nanoscale gap (stage 2 in Figure 1c). In such cases, the current trace exhibits characteristic step-like features or plateaus corresponding to the quantum conductance of the molecular junction. Upon further retraction of the tip, the molecule detaches from one electrode (stage 3 in Figure 1c). The collapse of the molecular junction (open gap) is accompanied by a sharp drop in current. In a typical dynamic BJ experiment, we collect thousands of current traces, and an automated selection process designed in LabVIEW selects the decays that show plateaus (formation of a molecular junction and accumulates them into 1D semi-logarithmic conductance histograms [26]. The same selection criteria to build all histograms are applied throughout all experimental series. The yield of decay curves showing plateaus that meet the selection criteria is typically around 25% of the total number of collected curves [43]. The data selection process results in 1D conductance histograms that show peaks above the tunnelling background baseline, providing an averaged single-molecule conductance value, G. G is defined as $G = I_{step}/V_{bias}$, where I_{step} is the current plateau and V_{bias} is the voltage applied between the two Au electrodes.

In the static approach (Figure 1d) [5], a fixed inter-electrode distance (electrode-electrode separation) is established between the STM tip and the substrate by setting a defined tunnelling current, I_t, to flow between them. When the gap has stabilized and shows a constant I_t versus time signal, the I_t feedback is turned off and the actual tunnelling current monitored (stage 1 in Figure 1d). A sudden current increase to I_m appears when a target molecule is spontaneously caught in the gap and transiently forms a molecular junction, lowering the resistance between the electrodes (stage 2 in Figure 1d) [44]. This sudden increase in current is commonly referred to as a *blink*. A current blink is characterized by its conductance value and by its lifetime. The (spontaneous) collapse of the junction (i.e., molecule detaches from one electrode) is evidenced by a sudden drop of the I_m to I_t [5,45,46] (stage 3 in Figure 1d). The number of individual blinks per fraction of time defines the accumulation yield and hundreds of them are plotted as 1D and 2D conductance histograms with a common x-axis time origin. All blinks are accumulated for evaluation without any selection using an automated process designed in LabVIEW. This process provides the averaged single-molecule conductance as a peak in the 1D histograms and prominent coloured regions of higher counts in the 2D blinking conductance-map histograms. The current corresponding to a trapped molecule (I_{blink}) is defined as $I_{blink} = I_m - I_t$. Conductance values G are extracted from I_{blink} divided by the V_{bias}.

The 1D semi-logarithmic conductance histogram retrieved from single-molecule dynamic BJ experiments in the *trans*-TMC molecules exhibits two broad, overlapping conductance features (Figure 2); 1.68×10^{-4} G_o (low conductance regime, LC) and 3.21×10^{-4} G_o (high conductance regime, HC), where $G_o = 2e^2/h = 77.47$ μS, the *h* Planck constant, the *e* elementary charge. The HC value is roughly twice of the LC value which might suggest the formation of multiple *trans*-TMC molecular junctions. However, several experimental observations rule out previous scenario: first, all the collected current traces show no correlation, displaying either HC or LC plateaus (Figure 2b), as opposed to current traces of junctions formed by multiple molecules that are characterized by the appearance of sequential plateau features due to consecutive molecular disconnections [47]. Second, junctions formed by multiple molecules also have a characteristic formation yield that decreases with the number of trapped molecules, i.e., the frequency of molecular junction formation decreases from the low conductive (few molecules) to the high conductive junctions (many molecules) [47,48]. The latter is the behaviour opposite to that observed in Figure 2a where the HC yield (27.7%) is significantly higher than the LC yield (17.4%). As such, we rule out that the HC values are related to a multiple molecule junction scenario and that the detection of the two current signatures must be a consequence of distinct single-molecule properties as discussed below.

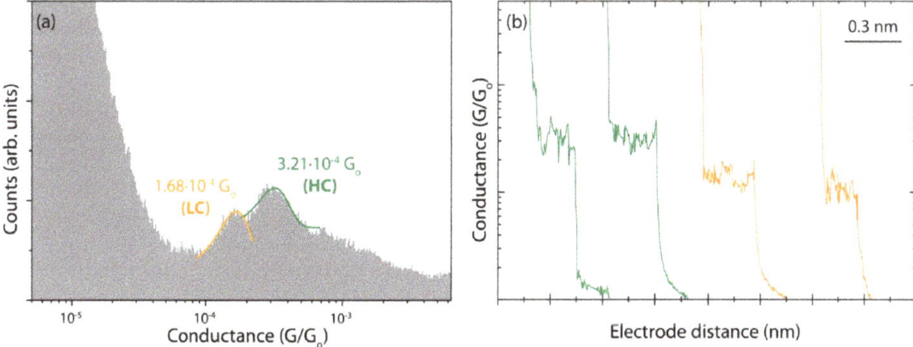

Figure 2. (a) Semi-logarithmic histogram from dynamic break junction (BJ) experiments for trans-TMC displaying a high conductance peak (HC, green traces) and a low conductance peak (LC, orange traces) fitted with Gaussians. (b) Representative individual current versus gap distance traces. V_{bias} = 20 mV.

The nanoscale inter-electrode distance of a molecular junction allows strong electric fields in the gap in the order of 10^8 to 10^9 V/m [5]. Such strong fields have been reported to promote single-molecule reactions of trapped molecules in an STM junction [3]. Given the cis-/trans- isomerization in carotenoids, we explore the possible electric field effect on the isomerization reaction in these systems [25,49–53]. The dynamic BJ approach can have detrimental effects when studying molecular configuration dynamics in a molecular junction due to the mechanical stress induced by the STM tip during retraction, which pulls on the molecule and can induce structural modifications [25,49–51]. Instead, the static BJ approach (Figure 1b) avoids the aforementioned drawbacks by employing a fixed inter-electrode distances also ensuring a constant gap EEF strength. The static BJ approach allows for a precise oriented EEF along the main junction axis. To this aim, we define three EEF of increasing magnitudes (low, intermediate, and high) by applying V_{bias} values of 7 mV, 65 mV and 130 mV. For a tip-substrate gap of approximately 3 nm (estimated length of the *trans*-TMC), the corresponding gap EEFs are ca. 2.3×10^{-3} V/nm, 2.2×10^{-2} V/nm, and 4.3×10^{-2} V/nm, respectively. Figure 3b–d shows the 2D blinking conductance-map histograms obtained for the three applied EEFs. In the low bias regime of 7 mV, the *trans*-TMC displays HC events only at ca. 3×10^{-4} G_o (Figure 3b). At intermediate and high V_{bias} of 65 and 130 mV, the *trans*-TMC 2D conductance-maps additionally exhibit a LC feature with blinks at ca. 1.5×10^{-4} G_o (Figure 3c,d). These results are in agreement with the dynamic BJ results and provide further evidence that the observed conductance features originate from two distinct molecular features in the junction. The blink lifetime is the highest for the low-bias regime, reaching 0.73 s, decreasing to 0.44 s for HC and 0.58 s scores for LC at intermediate V_{bias} and the shortest blink lifetimes, of 0.32 s for HC and of 0.42 s for LC, at high V_{bias}. The decrease in blink lifetime with increasing V_{bias} can be explained by a decrease in junction stability due to local heating effects and electromigration-induced mobility of metal atoms under higher EEFs. [54]. Note that at all the applied bias voltages, HC blinks display a 1.3 times longer lifetime as compared to LC blinks, which denotes an intrinsically higher junction stability in the case of HC junctions. The relative populations of the HC and LC junctions display a clear dependence on the applied EEFs. At low bias, all detected blinks correspond to the HC regime while at intermediate and high biases, the HC yield decreases significantly but LC features increase. At V_{bias} = 65 mV, the LC/HC yield ratio is nearly one to one, 48 /52, respectively. At V_{bias} = 130 mV, the LC/HC yield ratio increases to more than three, with the LC yield reaching 77 % of the total captured blinks. The observed behaviour in the blinking BJ approach is in accordance with the dynamic BJ results. In dynamic BJ experiments the varying gap distances produce overall higher EEFs even at the lower V_{bias} due to the transiently lower electrode–electrode gap distances (starting at direct electrode-electrode contact in the break-junction cycle, Figure 1c), which are sufficiently strong to induce the in situ *trans*-to-*cis* isomerisation and resulting in observed larger LC (*cis*-TMC) yields than in the blinking experiments. The observations for both BJ approaches suggest a conversion of HC to LC regimes in the presence of a sufficiently large V_{bias}, i.e., a sufficiently strong applied EEF.

To test the HC to LC switching hypothesis, we perform voltage-pulse experiments where a high bias voltage pulse of 300 mV is applied for 10 ms during the lifetime of a HC blink to induce the switching to LC (Appendix A, Figure A1). After the voltage pulse, the junction switches from HC to LC and stays at the characteristic LC value at the returning low V_{bias} of 30 mV prior to junction breakdown. Analogous voltage-jump experiments performed during LC blinks, show no conversion from LC to HC values. Both sets of experiments attest an irreversible process for the HC-to-LC conversion.

What is the nature of the two distinct V_{bias}-dependent conductance features for *trans*-TMC in Figure 3? The static BJ results rule out a mechanical origin of the observed conductance switching initially observed in the dynamic BJ experiments, and point toward the bias voltage as the driving force for the observed bimodal conductance behaviour. Guided by previous literature that showed isomerisation reactions of individual molecules tuned by voltage differences [52,55], we hypothesize that applying a V_{bias} to the TMC junction induces a *trans*-to-*cis* isomerisation of the molecule in the junction [52]. We have synthesized a homologous *cis* carotenoid (*cis*-TMC-syn) with a nearly identical structure (Figure 1b) and use it as a control test. The *cis*-TMC-syn control junctions show identical yield, conductance and blinking lifetime as the LC regime observed for the *trans*-TMC junctions (see Appendix A.4 for details). These control findings strongly support our hypothesis that the *trans*-TMC undergoes an in situ isomerisation induced by the applied EEF along the junction.

The observed irreversibility of the isomerisation can be explained by preferential molecule｜electrode coupling of the *cis*-TMC compared to *trans*-TMC [9]. We speculate that the *cis*-TMC structure offers an increased coupling between the distal phenyl ring and the Au electrode through van-der-Waals (π-) interactions [56,57]. As a consequence of the *cis*-TMC geometry adopted during the single-molecule contact, the phenyl ring comes closer and laterally faces the Au electrode [9], promoting a π-stacking interaction, in additional to the thiol bond [56–58], with interaction energies in the order of 20 kcal/mol [59]. The relative orientation of phenyl moiety with respect to the Au electrode and its close distance to the electrode surface is known to play an important role for stability and transport properties in single-molecule junctions [57]. Thus, the methyl-S–Au bond accompanied by phenyl–Au interactions provide the larger stability of the *cis*-TMC single-molecule junction relative to the *trans* isomer [60]. Such cooperative stabilisation effect can be expected to lead to longer junction lifetimes and non-reversibility of the in situ *trans*-to-*cis* isomerisation, in line with our observations. Despite the larger molecule | electrode coupling for *cis*-TMC, the overall conductance of the *cis*-TMC junction is dominated by the *gauche* defect that breaks the π-orbital electron pathway through the molecule [32], yielding a lower conductance.

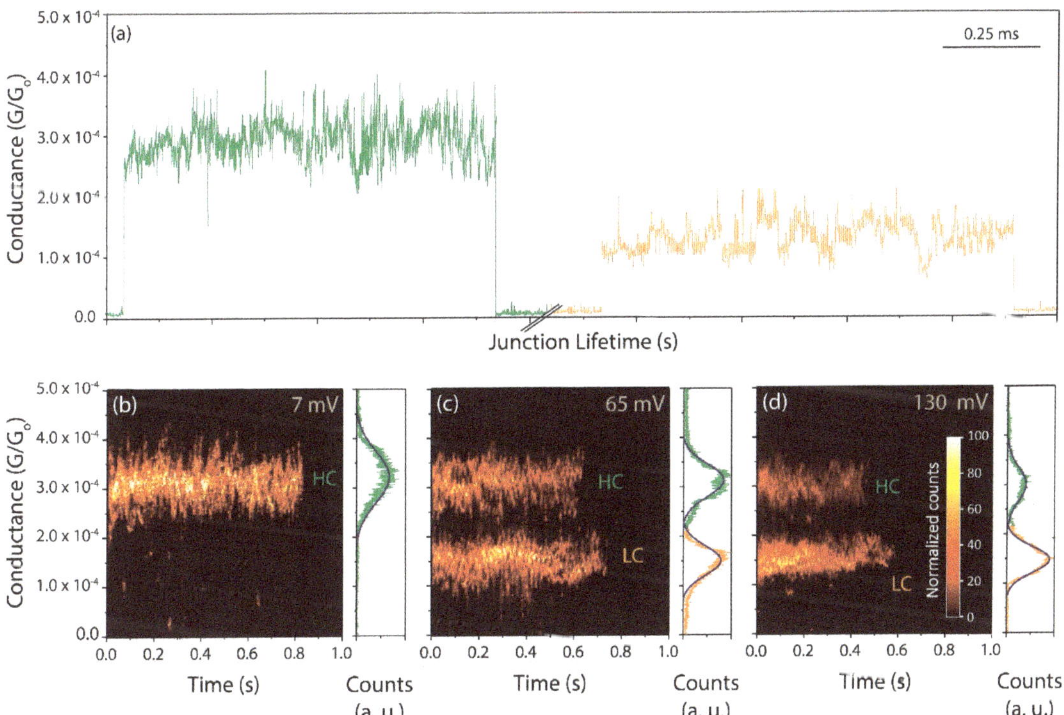

Figure 3. (**a**) Representative HC (green) and LC (orange) blinks of the trans-TMC junction formation in a typical static BJ experiment. 2D conductance maps built out of hundreds of trans-TMC blinking traces at V_{bias} = (**b**) 7 mV, (**c**) 65 mV, and (**d**) 130 mV at a fixed gap distance. Counts are normalized to a maximum of 100.

To corroborate the hypothesis of in situ EEF-induced *trans*-to-*cis* isomerisation in the single-TMC junction, we performed density functional theory (DFT) calculations to model the effect that an oriented EEF has on the *trans*-TMC molecule in the junction. Low-energy excited states are often the key to describe photoinduced isomerisation processes [61–63]. We then analysed how an electric field affects these energy levels via applied voltage bias. For the S_1 (first excited state), a dramatic change in the oscillator strength (spectroscopic transition probability) is observed for both the *cis*-TMC and *trans*-TMC conformers. Figure 4 shows (black line) that the $S_0 \rightarrow S_1$ transition is strongly forbidden in the absence of an electric field and is associated with delocalized orbitals extending along the molecule (see natural transition orbitals for the *trans*-TMC conformer in Figure 5). The addition of an EEF (blue and red curves in Figure 4) increases the oscillator strength dramatically for both *cis*-TMC and *trans*-TMC conformers (0° and 180° respectively along the X-axis rotation coordinate depicted in Figure 4) due to conjugation along the molecular chain, leading to large polarizability. In the case of *trans*-TMC, the enhancement of the oscillator strength is less dramatic than in the *cis*-TMC conformer (a factor of 12 instead of 600, Figure 4) because the transition to the S_1 state at zero field for *trans*-TMC is weak but not as forbidden as for *cis*-TMC. These differences are related to the reduction of symmetry associated with double bond rotation (see C_9–C_{10}

in Figure 1a), as illustrated in Figure A4. In *cis*-TMC case, natural transition orbitals [64], localized at each side of the double bond, indicate that the transition to S_1 will lead to the charge displacement along the charge transport direction, contributing to the current.

Calculated energy profiles of the TMC *trans*-to-*cis* isomerisation in the absence of an EEF for molecules in the ground state S_0, show expectedly large energy rotation barriers of ca. 48 (30) kcal/mol (Appendix A, Figure A3). These barriers are consistent with previous estimates of C=C double-bond rotation activation energies [65]. EEF, however, promotes the mixing of the S_1, S_2 and S_3 states, in line with the observed enhancement of the oscillator strength (the overlap matrix is given in the Table A1 of the Supplementary Information). For *trans*-TMC molecules, the oscillator strengths of the $S_0 \rightarrow S_1$ transition at fields on the order of 4.3×10^{-2} V/nm increases to 0.2. Therefore, a voltage difference in the order of a few hundreds of mV for an inter-electrode distance (nanogap) adapted to the molecular size, should be sufficiently strong to induce a mixing between the S_1 and S_2 states, suggesting that the S_1 state participates in the current-induced excited-state dynamics and promotes the isomerisation. In our case, DFT calculations indicate that a voltage bias is likely to enhance the contribution of the S_1 state to the electron dynamics, promoting the isomerisation process. The applied bias voltage does not need to overcome the double-bond rotation barrier to promote conformational switching. Travelling electrons will activate isomerisation pathways by populating the π^* orbitals connected with photo switching processes, as evidenced by the dramatic EEF- induced mixing of the key S_1 state.

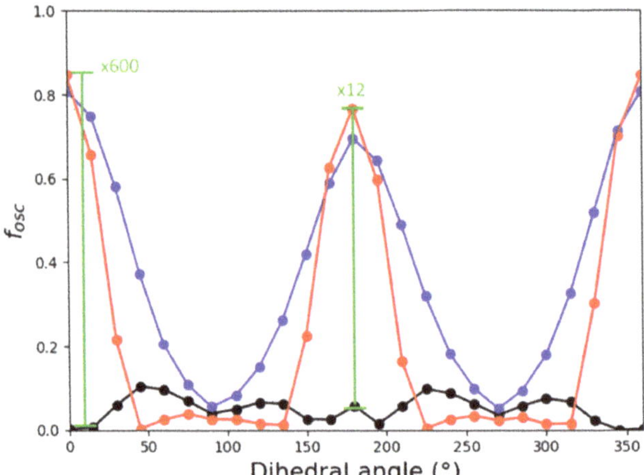

Figure 4. Oscillator strength for the $S_0 \rightarrow S_1$ transition. Black, blue, and red colours correspond to external electric fields (EEF) = 0, $+2.3 \times 10^{-4}$ a.u. and -2.3×10^{-4} a.u., respectively. Note the strength difference in the enhancement of the oscillator between trans-TMC and cis-TMC conformers (a factor of 12 and 600, respectively).

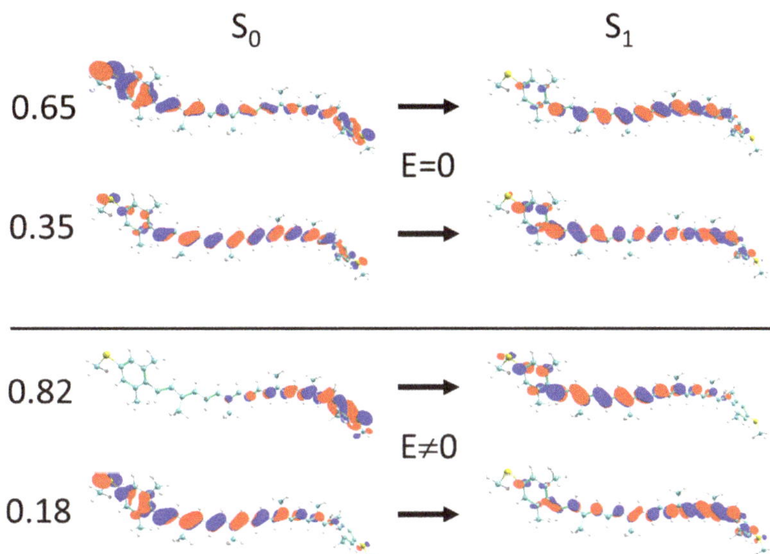

Figure 5. Orbitals for the $S_0 \rightarrow S_1$ natural transition of the trans-TMC in the absence (top) or presence (bottom) of an EEF (0° dihedral angle). Left numerical column indicates the weight for each orbital.

To summarize, we can rationalize all observed experimental results with the assignment of HC and LC values to *trans*-TMC and *cis*-TMC isomers, respectively:

Conductance. The *cis* isomer has a lower conductance value than the *trans* isomer. As previously reported [22], the breaking of the π-conjugation due to isomerisation lowers the transport efficiency along the polyene backbone [32].

Lifetime. The difference in lifetime between the HC blinks and the LC blinks reflects the different stability of each isomer in the molecular junction. The higher lifetimes of the LC blinks and the irreversibility of the *trans*-to-*cis* isomerisation can be attributed to a relatively higher stability of the *cis*-TMC isomer in the junction due to a more constrained configuration involving the terminal phenyl rings in the molecular anchoring to the electrodes.

Population vs. V_{bias}. The ratio of LC/HC blinks increases with increasing V_{bias}. DFT calculations suggest that the S_1 state remains spectroscopically dark at low fields, and therefore cannot promote electron transport. Thus, only high conductance associated with the *trans*-TMC isomer is observed. At high biases, the EEF induces an excited state mixing and hence promotes the *trans*-*cis* isomerisation, leading to an increasing number of (less conductive) cis conformers.

4. Conclusions

In the present study, an in situ electric field-dependent isomerisation in a single-molecule junction is demonstrated. Our experimental platform allows modulating the EEF by modifying the bias voltage magnitude to well-defined nano-gaps between two metal leads, while detecting single-molecule current events. By characterizing both a *trans*-TMC and a synthetically homologous *cis*-TMC version, we have univocally assigned molecular conductance features to each isomer. *Trans*-TMC is then ascribed to the molecular junction displaying average conductance values of ca. 3×10^{-4} Go, and the in situ generated *cis*-TMC is ascribed to molecular junctions average conductance values of ca. 1.6×10^{-4} Go. We show that controlled voltage pulses can be used to in situ isomerize the trapped polyene molecular backbone in the nanoscale molecular junction. Our combined experimental results and DFT calculations ascertain that a controlled isomerisation irreversibly occurs in the *trans*-TMC-based single-molecule junctions when converted into a more structurally stable *cis*-TMC-based junction. The isomerisation rate dependence on the EEF strength is supported by the increasing yield of junction formation of the *cis*-TMC form as a function of the EEF magnitude. Under high bias regimes (i.e., strong EEF), the *cis*-TMC LC feature dominates the conductance histogram. We conclude that the isomerisation from the *trans*-to the *cis*-TMC is promoted in a controlled way under a specific EEF strength in the single-molecule junctions. These findings have a special relevance in the field of electrocatalysis, since obtaining the *cis*-TMC isomer is unfeasible using traditional bulk synthetic routes due to steric constrains.

Author Contributions: Conceptualisation, A.C.A., I.D.-P., S.K.; Synthesis: S.K., methodology, A.C.A.; software, A.C.A., D.A. (Daniel Aravena); validation, C.S.Q., A.C.A., D.A. (Daniel Aravena)., I.D.-P.; formal analysis, C.S.Q., A.C.A., D.A. (Daniel Aravena), D.A. (Denis Andrienko), K.F.D., I.D.-P.; investigation, C.S.Q., A.C.A.; resources, A.C.A., D.A, S.K., I.D.-P.; data curation, A.C.A., D.A. (Daniel Aravena), K.F.D.; writing—original draft preparation, A.C.A., K.F.D.; writing—review and editing, C.S.Q., A.C.A., D.A. (Denis Andrienko), K.F.D., S.K., I.D.-P.; visualisation, A.C.A., D.A. (Daniel Aravena); supervision, A.C.A., K.F.D., D.A. (Daniel Aravena), S.K., I.D.-P.; project administration, A.C.A.; funding acquisition, K.F.D., A.C.A., S.K., I.D.-P. All authors have read and agreed to the published version of the manuscript.

Funding: I.D.-P. thanks the ERC (Fields4CAT-772391) for financial support. A.C.A. thanks European Union for a H2020-MSCA-IF-2018 Fellowship (TECh-MoDE). K.F.D. is grateful for generous funding through the "Plus 3" program of the Boehringer Ingelheim Foundation. S.K. appreciate National Research Foundation of Korea for a research grant (NRF-2020R1A2C1010724). D.A. thanks Powered@NLHPC; this research was partially supported by the supercomputing infrastructure of the NLHPC (ECM-02).

Institutional Review Board Statement: Not applicable.

Informed Consent Statement: Not applicable.

Data Availability Statement: All the data and additional information supporting the findings of this study are available from the corresponding authors upon reasonable request.

Conflicts of Interest: The authors declare no conflict of interest.

Appendix A

Appendix A.1. In Situ Irreversible High-Conductance (HC) to Low-Conductance (LC) Conversion

To induce HC-to-LC conversion in situ and to study its reversibility [6,52], a high-voltage pulse of 300 mV is applied for 10 ms during the lifetime of a detected HC blink, with the aim to switch the molecular junction to the LC regime (Figure A1a,b). Immediately after the pulse, at V_{bias} = 30 mV, the detected current decreases to the characteristic LC value (Figure A1c), attesting the V_{bias}-induced conversion of an individual trapped *trans*-TMC molecule. Eventually, the junction collapses spontaneously and the current suddenly drops to the I_t value. Equivalent experiments performed during LC blinks show constant I_m profiles independent of the bias pulse, without conversion from LC to HC values (see Figure A1d). Both sets of experiments point to an irreversible HC-to-LC conversion under the experimental conditions.

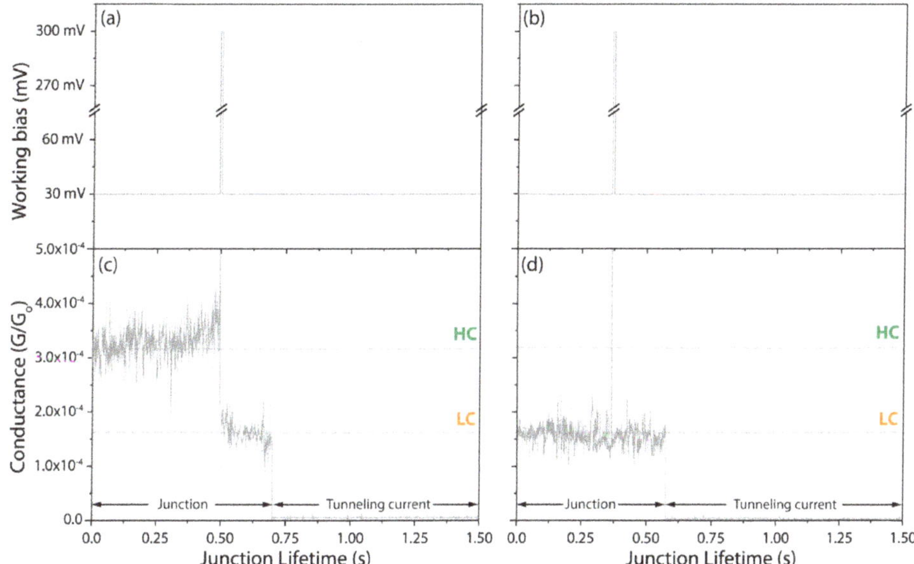

Figure A1. (**a**,**c**) V_{bias} versus time traces indicating the 10 ms short voltage pulse of 300 mV. (**b**,**d**) Conductance versus time traces corresponding to the voltage schemes depicted in (**a**,**c**) at a fixed inter-electrode distance of ca. 3 to 3.5 nm.

Appendix A.2. Energy Profile for trans-to-cis Rotation in the Absence of an Electric Field

Geometry optimisations for the *cis*-TMC and *trans*-TMC configurations reveal that both isomers present a similar energy, with a difference of 0.82 kcal/mol favouring the *cis*-TMC isomer. Of course, this energy difference is below the typical error of DFT methods, so we can consider both isomers to be of basically the same energy. Starting from the optimized *cis*-TMC isomer, the dihedral angle connecting both structures was rotated in $15°$ intervals (Figure A2). The ground state energy profile was insensitive to the addition of an electric field of $\pm 2.2 \times 10^{-4}$ atomic units (a.u.), i.e., 1.1×10^{8} V/nm, and curves are indistinguishable from the ground state in Figure A3a.

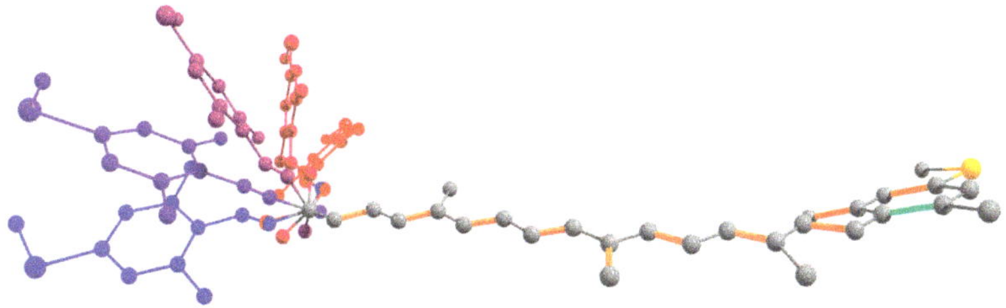

Figure A2. Superposition of structures with $\theta = 0°$ (blue), $45°$, $90°$, $135°$, $180°$ (red). Colour code: S, yellow; C: grey. Hydrogen atoms are omitted for clarity. The structures have not been reoptimized for each studied dihedral angle.

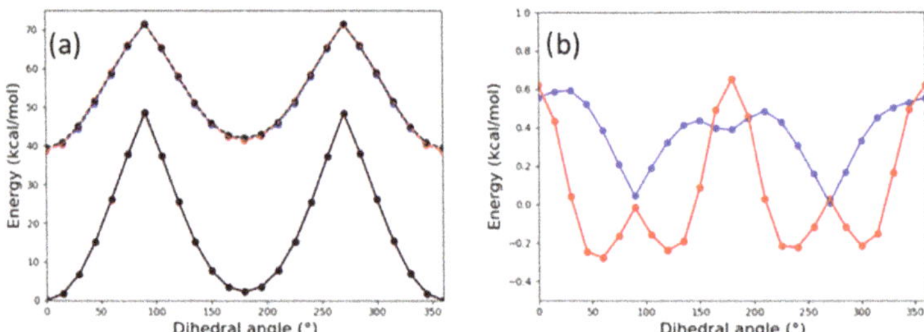

Figure A3. (**a**) Energy of the ground (solid lines) and first singlet excited (dashed) states along the rotation of the C–C double bond connecting cis-TMC and trans-TMC configurations. (**b**) Stabilisation of the S_1 state due to the inclusion of an EEF of 2.2×10^{-4} a.u. in the z direction. Blue and red colours correspond to positive and negative signs for the field.

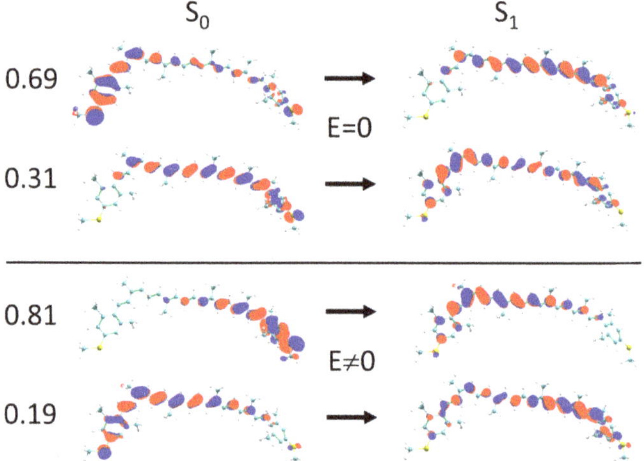

Figure A4. Natural transition orbitals [64] for the $S_0 \rightarrow S_1$ transition of the cis-TMC in the absence (top) or presence (bottom) of an EEF ($180°$ dihedral angle). The weight of each contribution is indicated at the left side.

Table A1. Overlap matrix for the ground state (GS) and 10 lowest singlet excited states (S_n) in the presence and absence of an EEF of $+2.2 \times 10^{-4}$ a.u. Larger positive and negative contributions are highlighted in blue and red, respectively.

cis-TMC isomer:

	GS	S_1	S_2	S_3	S_4	S_5	S_6	S_7	S_8	S_9	S_{10}
GS	0.9995	0.0004	0.0262	−0.0148	−0.0003	−0.0046	−0.0004	−0.0002	−0.0001	−0.0003	0.0001
S_1	−0.0112	0.8926	0.4447	0.0352	0.0272	0.0565	0.0076	−0.0126	0.0002	0.0043	0.0044
S_2	0.0236	0.4445	−0.8925	0.0177	−0.0617	0.0302	0.0000	0.0173	0.0012	0.0012	−0.0063
S_3	−0.0140	0.0333	0.0324	−0.8770	−0.4738	0.0048	−0.0302	0.0462	−0.0019	−0.0091	0.0282
S_4	0.0053	−0.0230	0.0590	0.4740	−0.8741	0.0295	−0.0677	−0.0331	−0.0067	−0.0221	−0.0137
S_5	0.0042	−0.0623	−0.0051	−0.0230	0.0285	0.9713	0.0093	−0.2210	0.0008	−0.0008	0.0487
S_6	−0.0003	0.0046	−0.0029	−0.0077	0.0506	−0.0064	−0.7369	−0.0596	−0.5787	0.3402	−0.0127
S_7	−0.0014	0.0138	−0.0203	−0.0528	0.0092	−0.2141	−0.0937	−0.9482	0.1290	−0.1580	−0.0343
S_8	−0.0001	0.0038	0.0041	0.0060	0.0582	0.0538	−0.6394	0.2106	0.4624	−0.5703	−0.0387
S_9	0.0002	−0.0023	−0.0001	0.0049	−0.0080	−0.0081	−0.1834	−0.0161	0.6555	0.7209	0.1284
S_{10}	−0.0002	−0.0016	0.0081	−0.0304	−0.0037	0.0520	0.0145	0.0141	0.0701	0.1171	−0.9886

Table A1. *Cont.*

trans-TMC isomer:

	GS	S_1	S_2	S_3	S_4	S_5	S_6	S_7	S_8	S_9	S_{10}
GS	0.9997	−0.0013	−0.0217	−0.0123	0.0029	−0.0052	0.0008	−0.0013	−0.0002	0.0009	0.0017
S_1	0.0101	−0.8381	0.5414	−0.0248	−0.0201	−0.0566	0.01	−0.007	0.0017	0.0023	−0.0008
S_2	−0.0192	−0.5417	−0.8386	−0.0041	0.0462	−0.0205	−0.0155	0.0062	−0.0006	−0.0016	−0.0057
S_3	0.0106	−0.0178	0.0302	0.9064	0.4184	0.0022	−0.0217	0.0244	−0.0037	−0.0004	0.0314
S_4	−0.007	0.0188	0.0404	−0.4185	0.9042	−0.0309	0.0375	0.0506	−0.0018	−0.0149	−0.0064
S_5	−0.0053	0.0568	−0.0184	0.0222	−0.0287	−0.9782	0.189	−0.0427	−0.0001	0.0055	0.0241
S_6	−0.0013	0.006	0.0045	−0.019	0.0511	−0.0663	−0.5116	−0.7544	−0.3966	−0.0344	0.0567
S_7	−0.0002	−0.0101	−0.0151	0.0287	0.0066	0.1766	0.829	−0.419	−0.3031	−0.0968	−0.0596
S_8	−0.0009	0.0002	−0.0059	0.0014	0.0336	0.0388	0.0798	−0.4876	0.8005	0.3317	0.0489
S_9	−0.0016	−0.0022	−0.0036	−0.0167	0.0005	0.0221	0.0652	0.1119	−0.3153	0.8471	0.4066
S_{10}	−0.0014	−0.0049	−0.0045	−0.0245	−0.0119	0.0293	0.0488	−0.0036	0.1032	−0.4018	0.9077

Appendix A.3. cis-TMC-syn (Synthetic) Carotenoid Control Experiments

As a control experiment for the *trans*-to-*cis* isomerisation process, we employ a homologous *cis*-TMC isomer control molecule, the *cis*-TMC-syn (see Figure 1b). The control molecule possesses an almost identical structure to *cis*-TMC and thus represents a suitable molecular reference to assign the LC feature in (Figure 3) to the cis-isomer. The missing methyl substituents on the phenyl rings on the carotenoid control compared to TMC are not expected to affect significantly the characteristic conductance value of the conjugated carotenoid wire, since CH_3- and H-terminal groups exhibit comparable electron-donating character [26].

Unlike for *trans*-TMC that shows two discrete LC and HC regimes in the dynamic approach, the dynamic BJ *cis*-TMC-syn experiments exhibit a single (low) conductance value (Figure A5a). *cis*-TMC-syn yield of 20.3 % and the conductance value of 1.77×10^{-3} G_o are nearly identical to the analogous concepts for *trans*-TMC junctions: a yield of 17.4 % and a LC value of 1.68×10^{-3} G_o. as Figure 2a shows.

Figure A5c–e displays the 2D conductance maps of the *cis*-TMC-syn at the different applied EEF. Unlike for *trans*-TMC, *cis*-TMC-syn junctions show a unique and constant current signature at all three applied V_{bias} regimes. The molecular conductance value extracted from the blinking approach of 1.60×10^{-3} G_o is in good agreement with the conductance value obtained from the dynamic approach. The extracted conductance values from the 2D blinking histograms of *cis*-TMC-syn junctions are statistically identical to those of the LC regimes measured in the *trans*-TMC at the same V_{bias} (Figure A6a, blue and orange plots).

Another remarkable similitude between *cis*-TMC-syn and LC *trans*-TMC junctions is the observed blinking lifetime that are correlated under equivalent bias regimes (see Figure A6b, blue and orange, respectively). For both measured species, the blinking lifetime is ca. a factor 1.3 larger than that of the HC *trans*-TMC junctions.

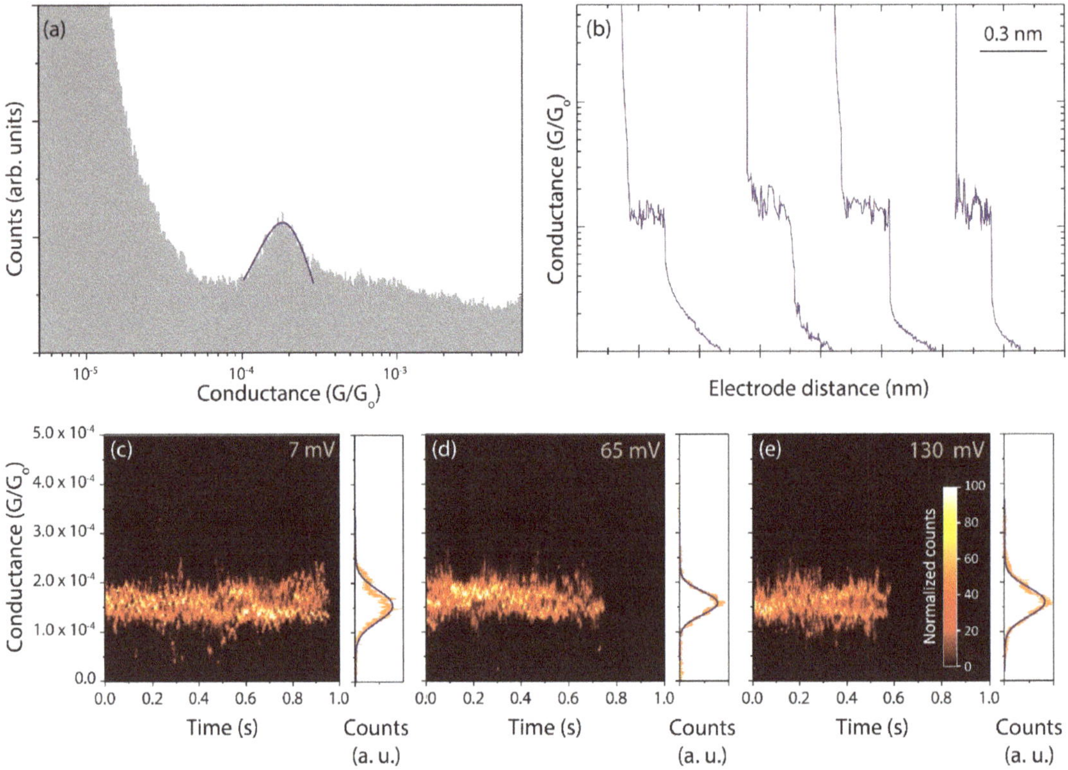

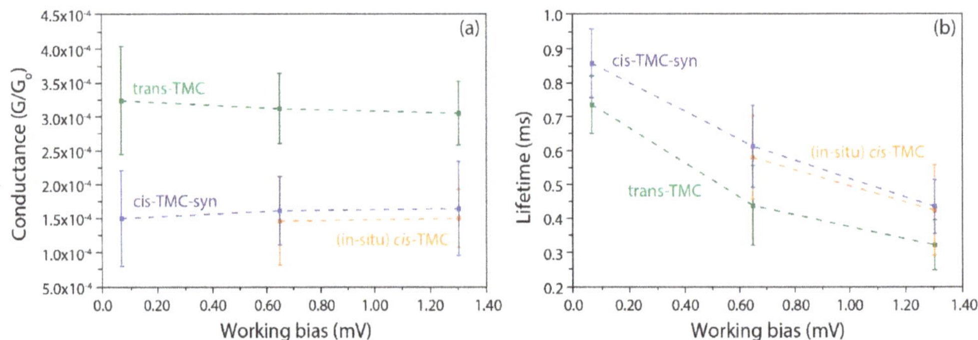

Figure A5. (**a**) Semi-logarithmic histogram from dynamic experiments for cis-TMC-syn control molecule displaying a single peak of LC fitted with a Gaussian. (**b**) Representative individual current versus gap distance traces. V_{bias} = 20 mV. (**c–e**) 2D conductance maps of the cis-TMC-syn control molecule blinking experiments V_{bias} = (**b**) 7 mV, (**c**) 65 mV and (**d**) 130 mV at a fixed gap distance. V_{bias} was randomly changed for each set of experiments for a total time of 5h for each bias regime. Counts are normalized to a maximum of 100.

Figure A6. (**a**) Molecular conductance and (**b**) blinking lifetimes for trans-TMC HC (green) and LC (orange) as well as for cis-TMC-syn (blue) junctions for the three employed V_{bias} values. Conductance and standard deviation values are extracted with Gaussian fits of the peaks in 1D blinking histograms of Figures 3 and A5.

Appendix A.4. Synthetic Details

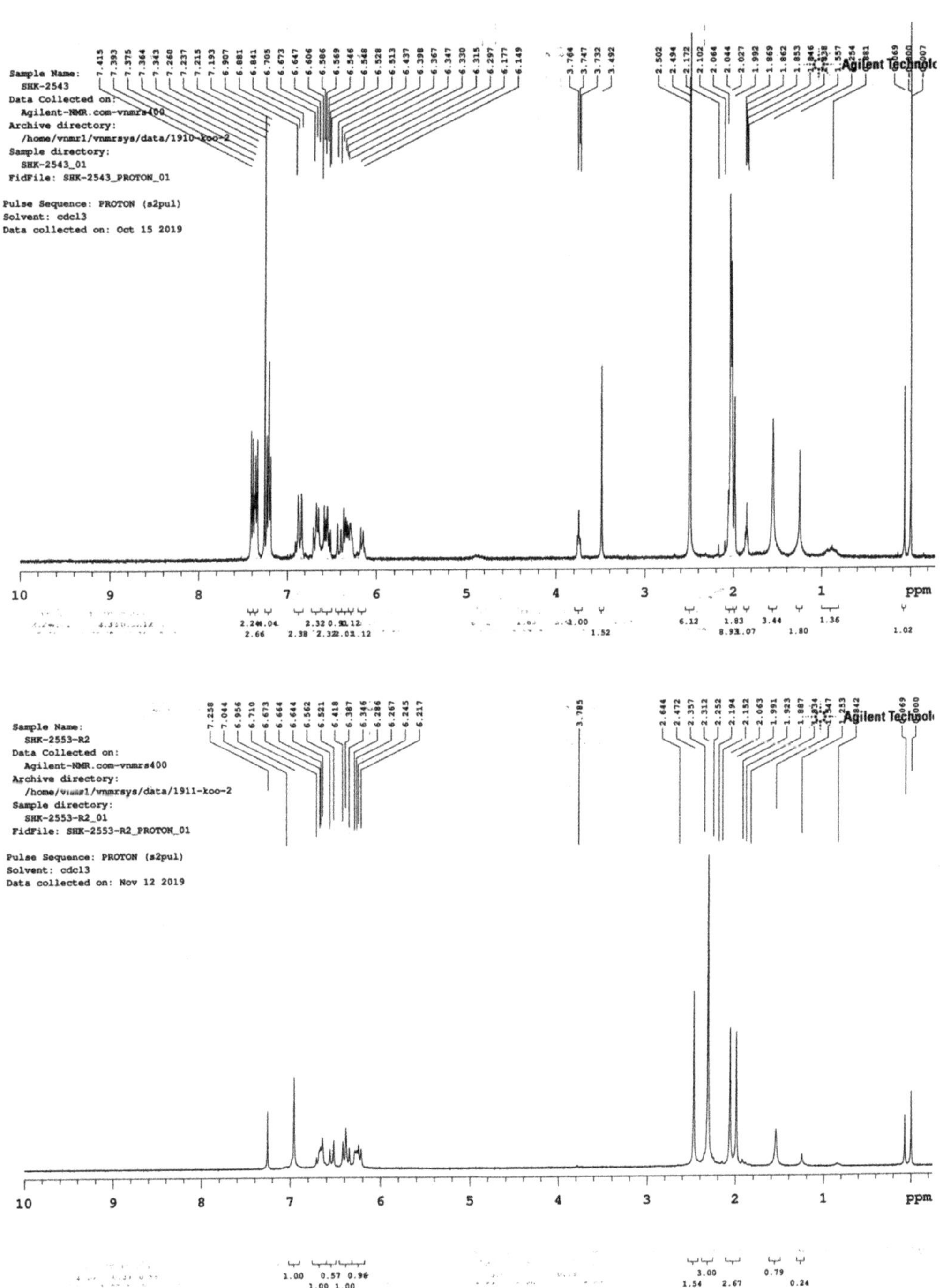

Figure A7. ^{1}H nuclear magnetic resonance (NMR) spectrum for *cis*-TMC-syn (top) and for *trans*-TMC (bottom).

References

1. Nair, V.; Muñoz-Batista, M.J.; Fernández-García, M.; Luque, R.; Colmenares, J.C. Thermo-Photocatalysis: Environmental and Energy Applications. *ChemSusChem* **2019**, *12*, 2098–2116. [CrossRef]
2. Shaik, S.; Danovich, D.; Joy, J.; Wang, Z.; Stuyver, T. Electric-Field Mediated Chemistry: Uncovering and Exploiting the Potential of (Oriented) Electric Fields to Exert Chemical Catalysis and Reaction Control. *J. Am. Chem. Soc.* **2020**, *142*, 12551–12562. [CrossRef]
3. Tang, C.; Zheng, J.; Ye, Y.; Liu, J.; Chen, L.; Yan, Z.; Chen, Z.; Chen, L.; Huang, X.; Bai, J.; et al. Electric-Field-Induced Connectivity Switching in Single-Molecule Junctions. *Science* **2020**, *23*, 100770. [CrossRef]
4. Gorin, C.F.; Beh, E.S.; Kanan, M.W. An Electric Field–Induced Change in the Selectivity of a Metal Oxide–Catalyzed Epoxide Rearrangement. *J. Am. Chem. Soc.* **2011**, *134*, 186–189. [CrossRef] [PubMed]
5. Aragonès, A.C.; Haworth, N.L.; Darwish, N.; Ciampi, S.; Bloomfield, N.J.; Wallace, G.G.; Diez-Perez, I.; Coote, M.L. Electrostatic catalysis of a Diels–Alder reaction. *Nat. Cell Biol.* **2016**, *531*, 88–91. [CrossRef] [PubMed]
6. Morgenstern, K. Isomerization Reactions on Single Adsorbed Molecules. *Accounts Chem. Res.* **2009**, *42*, 213–223. [CrossRef] [PubMed]
7. Kumar, A.S.; Ye, T.; Takami, T.; Yu, B.-C.; Flatt, A.K.; Tour, J.M.; Weiss, P.S. Reversible Photo-Switching of Single Azobenzene Molecules in Controlled Nanoscale Environments. *Nano Lett.* **2008**, *8*, 1644–1648. [CrossRef] [PubMed]
8. Mativetsky, J.M.; Pace, G.; Elbing, M.; Rampi, M.A.; Mayor, M.; Samorì, P. Azobenzenes as Light-Controlled Molecular Electronic Switches in Nanoscale Metal–Molecule–Metal Junctions. *J. Am. Chem. Soc.* **2008**, *130*, 9192–9193. [CrossRef]
9. Martin, S.; Haiss, W.; Higgins, S.J.; Nichols, R.J. The Impact of E–Z Photo-Isomerization on Single Molecular Conductance. *Nano Lett.* **2010**, *10*, 2019–2023. [CrossRef]
10. Donhauser, Z.J. Conductance Switching in Single Molecules Through Conformational Changes. *Science* **2001**, *292*, 2303–2307. [CrossRef] [PubMed]
11. Ramachandran, G.K.; Tomfohr, J.K.; Li, J.; Sankey, O.F.; Zarate, X.; Primak, A.; Terazono, Y.; Moore, T.A.; Moore, A.L.; Gust, D.; et al. Electron Transport Properties of a Carotene Molecule in a Metal–(Single Molecule)–Metal Junction. *J. Phys. Chem. B* **2003**, *107*, 6162–6169. [CrossRef]
12. Venkataraman, L.; Klare, J.E.; Nuckolls, C.; Hybertsen, M.S.; Steigerwald, M.L. Dependence of single-molecule junction conductance on molecular conformation. *Nat. Cell Biol.* **2006**, *442*, 904–907. [CrossRef] [PubMed]
13. Tao, N.J. Electron transport in molecular junctions. *Nat. Nanotechnol.* **2006**, *1*, 173–181. [CrossRef]
14. Cai, Z.-L.; Crossley, M.J.; Reimers, J.R.; Kobayashi, A.R.; Amos, R.D. Density Functional Theory for Charge Transfer: The Nature of the N-Bands of Porphyrins and Chlorophylls Revealed through CAM-B3LYP, CASPT2, and SAC-CI Calculations. *J. Phys. Chem. B* **2006**, *110*, 15624–15632. [CrossRef]
15. Leary, E.; Roche, C.; Jiang, H.-W.; Grace, I.; González, M.T.; Rubio-Bollinger, G.; Romero-Muñiz, C.; Xiong, Y.; Al-Galiby, Q.; Noori, M.; et al. Detecting Mechanochemical Atropisomerization within an STM Break Junction. *J. Am. Chem. Soc.* **2018**, *140*, 710–718. [CrossRef]
16. Li, J.; Tomfohr, J.K.; Sankey, O.F. Theoretical study of carotene as a molecular wire. *Phys. E Low Dimens. Syst. Nanostructures* **2003**, *19*, 133–138. [CrossRef]
17. Del Valle, M.; Gutierrez, R.; Tejedor, C.; Cuniberti, G. Tuning the conductance of a molecular switch. *Nat. Nanotechnol.* **2007**, *2*, 176–179. [CrossRef]
18. Dhivya, G.; Nagarajan, V.; Chandiramouli, R. First-principles studies on switching properties of azobenzene based molecular device. *Chem. Phys. Lett.* **2016**, *660*, 27–32. [CrossRef]
19. Li, C.; Pobelov, I.; Wandlowski, T.; Bagrets, A.; Arnold, A.A.; Evers, F. Charge Transport in Single Au | Alkanedithiol | Au Junctions: Coordination Geometries and Conformational Degrees of Freedom. *J. Am. Chem. Soc.* **2008**, *130*, 318–326. [CrossRef]
20. Cao, Y.; Dong, S.; Liu, S.; Liu, Z.; Guo, X. Toward Functional Molecular Devices Based on Graphene-Molecule Junctions. *Angew. Chem. Int. Ed.* **2013**, *52*, 3906–3910. [CrossRef]
21. Sotthewes, K.; Geskin, V.; Heimbuch, R.; Kumar, A.; Zandvliet, H.J.W. Research Update: Molecular electronics: The single-molecule switch and transistor. *APL Mater.* **2014**, *2*, 10701. [CrossRef]
22. Kim, M.; Jung, H.; Aragonès, A.C.; Diez-Perez, I.; Ahn, K.-H.; Chung, W.-J.; Kim, D.; Koo, S. Role of Ring Ortho Substituents on the Configuration of Carotenoid Polyene Chains. *Org. Lett.* **2018**, *20*, 493–496. [CrossRef]
23. Schwarz, F.; Lörtscher, E. Break-junctions for investigating transport at the molecular scale. *J. Phys. Condens. Matter* **2014**, *26*, 474201. [CrossRef] [PubMed]
24. Komoto, Y.; Fujii, S.; Iwane, M.; Kiguchi, M. Single-molecule junctions for molecular electronics. *J. Mater. Chem. C* **2016**, *4*, 8842–8858. [CrossRef]
25. Stefani, D.; Perrin, M.; Gutiérrez-Cerón, C.; Aragonès, A.C.; Labra-Muñoz, J.; Carrasco, R.D.C.; Matsushita, Y.; Futera, Z.; Labuta, J.; Ngo, T.H.; et al. Mechanical Tuning of Through-Molecule Conductance in a Conjugated Calix[4]pyrrole. *Chemistry* **2018**, *3*, 6473–6478. [CrossRef]
26. Aragonès, A.C.; Darwish, N.; Im, J.; Lim, B.; Choi, J.; Koo, S.; Díez-Pérez, I. Fine-Tuning of Single-Molecule Conductance by Tweaking Both Electronic Structure and Conformation of Side Substituents. *Chem. A Eur. J.* **2015**, *21*, 7716–7720. [CrossRef]
27. Chen, F.; Tao, N.J. Electron Transport in Single Molecules: From Benzene to Graphene. *Accounts Chem. Res.* **2009**, *42*, 429–438. [CrossRef]

28. Li, X.; He, J.; Hihath, J.; Xu, B.; Lindsay, S.M.; Tao, N. Conductance of Single Alkanedithiols: Conduction Mechanism and Effect of Molecule–Electrode Contacts. *J. Am. Chem. Soc.* **2006**, *128*, 2135–2141. [CrossRef]
29. Kushmerick, J.G.; Holt, D.B.; Pollack, S.K.; Ratner, M.A.; Yang, J.C.; Schull, T.L.; Naciri, J.; Moore, M.H.; Shashidhar, R. Effect of Bond-Length Alternation in Molecular Wires. *J. Am. Chem. Soc.* **2002**, *124*, 10654–10655. [CrossRef] [PubMed]
30. Tracewell, C.A.; Cua, A.; Stewart, D.H.; Bocian, D.F.; Brudvig, G.W. Characterization of Carotenoid and Chlorophyll Photooxidation in Photosystem II. *Biochemitry* **2001**, *40*, 193–203. [CrossRef] [PubMed]
31. Tracewell, C.A.; Brudvig, G.W. Multiple Redox-Active Chlorophylls in the Secondary Electron-Transfer Pathways of Oxygen-Evolving Photosystem II. *Biochemitry* **2008**, *47*, 11559–11572. [CrossRef] [PubMed]
32. Garner, M.H.; Solomon, G.C. Simultaneous Suppression of π- and σ-Transmission in π-Conjugated Molecules. *J. Phys. Chem. Lett.* **2020**, *11*, 7400–7406. [CrossRef] [PubMed]
33. Hunter, J.D. Matplotlib: A 2D Graphics Environment. *Comput. Sci. Eng.* **2007**, *9*, 90–95. [CrossRef]
34. Neese, F. Software update: The ORCA program system, version 4.0. *Wiley Interdiscip. Rev. Comput. Mol. Sci.* **2018**, *8*, 8. [CrossRef]
35. Runge, E.; Gross, E.K.U. Density-Functional Theory for Time-Dependent Systems. *Phys. Rev. Lett.* **1984**, *52*, 997–1000. [CrossRef]
36. Becke, A.D. Density-functional exchange-energy approximation with correct asymptotic behavior. *Phys. Rev. A* **1988**, *38*, 3098–3100. [CrossRef]
37. Perdew, J.P. Density-functional approximation for the correlation energy of the inhomogeneous electron gas. *Phys. Rev. B* **1986**, *33*, 8822–8824. [CrossRef]
38. Weigend, F.; Ahlrichs, R. Balanced basis sets of split valence, triple zeta valence and quadruple zeta valence quality for H to Rn: Design and assessment of accuracy. *Phys. Chem. Chem. Phys.* **2005**, *7*, 3297–3305. [CrossRef]
39. Weigend, F. Accurate Coulomb-fitting basis sets for H to Rn. *Phys. Chem. Chem. Phys.* **2006**, *8*, 1057–1065. [CrossRef]
40. Grimme, S.; Antony, J.; Ehrlich, S.; Krieg, H. A consistent and accurate ab initio parametrization of density functional dispersion correction (DFT-D) for the 94 elements H-Pu. *J. Chem. Phys.* **2010**, *132*, 154104. [CrossRef]
41. Grimme, S.; Ehrlich, S.; Goerigk, L. Effect of the damping function in dispersion corrected density functional theory. *J. Comput. Chem.* **2011**, *32*, 1456–1465. [CrossRef]
42. Plasser, F.; Ruckenbauer, M.; Mai, S.; Oppel, M.; Marquetand, P.; González, L. Efficient and Flexible Computation of Many-Electron Wave Function Overlaps. *J. Chem. Theory Comput.* **2016**, *12*, 1207–1219. [CrossRef] [PubMed]
43. Chen, F.; Li, X.; Hihath, J.; Huang, Z.; Tao, N. Effect of Anchoring Groups on Single-Molecule Conductance: Comparative Study of Thiol-, Amine-, and Carboxylic-Acid-Terminated Molecules. *J. Am. Chem. Soc.* **2006**, *128*, 15874–15881. [CrossRef] [PubMed]
44. Haiss, W.; Nichols, R.J.; Van Zalinge, H.; Higgins, S.J.; Bethell, D.; Schiffrin, D.J. Measurement of single molecule conductivity using the spontaneous formation of molecular wires. *Phys. Chem. Chem. Phys.* **2004**, *6*, 4330–4337. [CrossRef]
45. Pla-Vilanova, P.; Aragonès, A.C.; Ciampi, S.; Sanz, F.; Darwish, N.; Diez-Perez, I. The spontaneous formation of single-molecule junctions via terminal alkynes. *Nanotechnology* **2015**, *26*, 381001. [CrossRef] [PubMed]
46. Aragonès, A.C.; Darwish, N.; Ciampi, S.; Sanz, F.; Gooding, J.J.; Díez-Pérez, I. Single-molecule electrical contacts on silicon electrodes under ambient conditions. *Nat. Commun.* **2017**, *8*, 15056. [CrossRef]
47. Xu, B. Measurement of Single-Molecule Resistance by Repeated Formation of Molecular Junctions. *Science* **2003**, *301*, 1221–1223. [CrossRef] [PubMed]
48. Li, Z.; Han, B.; Meszaros, G.; Pobelov, I.; Wandlowski, T.; Błaszczyk, A.; Mayor, M. Two-dimensional assembly and local redox-activity of molecular hybrid structures in an electrochemical environment. *Faraday Discuss.* **2005**, *131*, 121–143. [CrossRef]
49. Bruot, C.; Hihath, J.; Tao, N. Mechanically controlled molecular orbital alignment in single molecule junctions. *Nat. Nanotechnol.* **2011**, *7*, 35–40. [CrossRef] [PubMed]
50. Inatomi, J.; Fujii, S.; Marqués-González, S.; Masai, H.; Tsuji, Y.; Terao, J.; Kiguchi, M. Effect of Mechanical Strain on Electric Conductance of Molecular Junctions. *J. Phys. Chem. C* **2015**, *119*, 19452–19457. [CrossRef]
51. Qi, J.; Gao, Y.; Jia, H.; Richter, M.; Huang, L.; Cao, Y.; Yang, H.; Zheng, Q.; Berger, R.; Liu, J.; et al. Force-Activated Isomerization of a Single Molecule. *J. Am. Chem. Soc.* **2020**, *142*, 10673–10680. [CrossRef]
52. Alemani, M.; Peters, M.V.; Hecht, S.; Rieder, K.-H.; Moresco, A.F.; Grill, L. Electric Field-Induced Isomerization of Azobenzene by STM. *J. Am. Chem. Soc.* **2006**, *128*, 14446–14447. [CrossRef] [PubMed]
53. Zang, Y.; Zou, Q.; Fu, T.; Ng, F.; Fowler, B.; Yang, J.; Li, H.; Steigerwald, M.L.; Nuckolls, C.; Venkataraman, L. Directing isomerization reactions of cumulenes with electric fields. *Nat. Commun.* **2019**, *10*, 1–7. [CrossRef] [PubMed]
54. Huang, Z.; Chen, F.; D'Agosta, R.; Bennett, P.A.; Di Ventra, M.; Tao, N. Local ionic and electron heating in single-molecule junctions. *Nat. Nanotechnol.* **2007**, *2*, 698–703. [CrossRef] [PubMed]
55. Blum, A.S.; Kushmerick, J.G.; Long, D.P.; Patterson, C.H.; Yang, J.C.; Henderson, J.C.; Yao, Y.; Tour, J.M.; Shashidhar, R.; Ratna, B.R. Molecularly inherent voltage-controlled conductance switching. *Nat. Mater.* **2005**, *4*, 167–172. [CrossRef] [PubMed]
56. Diez-Perez, I.; Hihath, J.; Hines, T.; Wang, Z.-S.; Zhou, G.; Müllen, K.; Tao, N. Controlling single-molecule conductance through lateral coupling of π orbitals. *Nat. Nanotechnol.* **2011**, *6*, 226–231. [CrossRef]
57. Kitaguchi, Y.; Habuka, S.; Okuyama, H.; Hatta, S.; Aruga, T.; Frederiksen, T.; Paulsson, M.; Ueba, H. Controlling single-molecule junction conductance by molecular interactions. *Sci. Rep.* **2015**, *5*, 11796. [CrossRef]
58. Yoshida, K.; Pobelov, I.V.; Manrique, D.Z.; Pope, T.; Mészáros, G.; Gulcur, M.; Bryce, M.R.; Lambert, C.J.; Wandlowski, T. Correlation of breaking forces, conductances and geometries of molecular junctions. *Sci. Rep.* **2015**, *5*, srep09002. [CrossRef]

59. Reckien, W.; Eggers, M.; Bredow, T. Theoretical study of the adsorption of benzene on coinage metals. *Beilstein J. Org. Chem.* **2014**, *10*, 1775–1784. [CrossRef]
60. Meisner, J.S.; Ahn, S.; Aradhya, S.V.; Krikorian, M.; Parameswaran, R.; Steigerwald, M.; Venkataraman, L.; Nuckolls, C. Importance of Direct Metal−π Coupling in Electronic Transport Through Conjugated Single-Molecule Junctions. *J. Am. Chem. Soc.* **2012**, *134*, 20440–20445. [CrossRef]
61. Roke, D.; Wezenberg, S.J.; Feringa, B.L. Molecular rotary motors: Unidirectional motion around double bonds. *Proc. Natl. Acad. Sci. USA* **2018**, *115*, 9423–9431. [CrossRef] [PubMed]
62. Hall, C.R.; Conyard, J.; Heisler, I.A.; Jones, G.A.; Frost, J.; Browne, W.R.; Feringa, B.L.; Meech, S.R. Ultrafast Dynamics in Light-Driven Molecular Rotary Motors Probed by Femtosecond Stimulated Raman Spectroscopy. *J. Am. Chem. Soc.* **2017**, *139*, 7408–7414. [CrossRef]
63. Pang, X.; Cui, X.; Hu, D.; Jiang, C.; Zhao, D.; Lan, Z.; Li, F. "Watching" the Dark State in Ultrafast Nonadiabatic Photoisomerization Process of a Light-Driven Molecular Rotary Motor. *J. Phys. Chem. A* **2017**, *121*, 1240–1249. [CrossRef] [PubMed]
64. Martin, R.L. Natural transition orbitals. *J. Chem. Phys.* **2003**, *118*, 4775–4777. [CrossRef]
65. Kazaryan, A.; Lan, Z.; Schäfer, L.V.; Thiel, W.; Filatov, M. Surface Hopping Excited-State Dynamics Study of the Photoisomerization of a Light-Driven Fluorene Molecular Rotary Motor. *J. Chem. Theory Comput.* **2011**, *7*, 2189–2199. [CrossRef] [PubMed]

Communication

The Effect of Anchor Group on the Phonon Thermal Conductance of Single Molecule Junctions

Mohammed D. Noori [1,2], Sara Sangtarash [1] and Hatef Sadeghi [1,*]

[1]   Device Modelling Group, School of Engineering, University of Warwick, Coventry CV4 7AL, UK; mdn.noori@sci.utq.edu.iq (M.D.N.); Sara.Sangtarash@warwick.ac.uk (S.S.)

[2]   Department of Physics, College of Sciences, University of Thi-Qar, Thi-Qar 64001, Iraq

*   Correspondence: Hatef.Sadeghi@warwick.ac.uk

Abstract: There is a worldwide race to convert waste heat to useful energy using thermoelectric materials. Molecules are attractive candidates for thermoelectricity because they can be synthesised with the atomic precision, and intriguing properties due to quantum effects such as quantum interference can be induced at room temperature. Molecules are also expected to show a low thermal conductance that is needed to enhance the performance of thermoelectric materials. Recently, the technological challenge of measuring the thermal conductance of single molecules was overcome. Therefore, it is timely to develop strategies to reduce their thermal conductance for high performance thermoelectricity. In this paper and for the first time, we exploit systematically the effect of anchor groups on the phonon thermal conductance of oligo (phenylene ethynylene) (OPE3) molecules connected to gold electrodes via pyridyl, thiol, methyl sulphide and carbodithioate anchor groups. We show that thermal conductance is affected significantly by the choice of anchor group. The lowest and highest thermal conductances were obtained in the OPE3 with methyl sulphide and carbodithioate anchor groups, respectively. The thermal conductance of OPE3 with thiol anchor was higher than that with methyl sulphide but lower than the OPE3 with pyridyl anchor group.

Keywords: molecular electronics; thermoelectricity; phonon; thermal conductance; OPE3; anchor groups; pyridyl; thiol; methyl sulphide; carbodithioate

Citation: Noori, M.D.; Sangtarash, S.; Sadeghi, H. The Effect of Anchor Group on the Phonon Thermal Conductance of Single Molecule Junctions. *Appl. Sci.* **2021**, *11*, 1066. https://doi.org/10.3390/app11031066

Academic Editor: Linda Angela Zotti
Received: 26 December 2020
Accepted: 21 January 2021
Published: 25 January 2021

Publisher's Note: MDPI stays neutral with regard to jurisdictional claims in published maps and institutional affiliations.

1. Introduction

Currently, nearly 10% of the world's electricity is used by computers and the internet and converted to heat. This waste heat could be used to generate electricity economically, provided materials with a high thermoelectric efficiency could be identified [1]. The demand for new thermoelectric materials has led to a worldwide race to develop materials with a high thermoelectric efficiency [2–11]. The efficiency of a thermoelectric device is inversely proportional to its thermal conductance $\kappa = \kappa_p + \kappa_e$ due to electrons (κ_e) and phonons (κ_p) [12]. Therefore, low-thermal-conductance materials are needed for an efficient conversion of heat into electricity. The state of the art thermoelectric figure of merit (ZT) was found in inorganic materials, e.g., 2.2 at high temperatures 900 K [13]. These inorganic materials are toxic, and their global supply is limited. Therefore, organic materials are now being considered [14].

Molecules are attractive candidates for thermoelectricity because their structure can be modified with atomic precision and desirable properties can be induced by the engineering of their structure [15,16]. They are also expected to show intriguing properties, such as room temperature quantum and phonon interference, that can be used to simultaneously increase their electrical conductance and Seebeck coefficient and to suppress their thermal conductance [17,18].

Just recently the technological challenge of measuring the thermal conductance of single molecules was overcome [19,20]. This opens new avenues to study the thermoelectric efficiency of single molecules [21,22]. To optimize molecular junctions for a maximum efficiency, strategies to increase their electrical conductance and Seebeck coefficient simultaneously and to suppress their thermal conductance should be developed. So far, thermal conductance of a few molecules, including C60 [23], alkanes [19,24], OPE2 derivatives [17], OPE3 [19,23], Benzene [25,26], Oligoyne [24], biphenyl-dithiol [17], bipyridyl and its radical counterpart [18], between gold electrodes has been calculated. Among these, thermal conductance of alkanes [19,20] and OPE3 [27] molecules with thiol anchor were measured. Table 1 shows a summary of room temperature thermal conductance calculations due to electrons (κ_e) and phonons (κ_p) of single molecules between gold electrodes using density functional theory (DFT) combined with the non-equilibrium Green's function (NEGF) method for transport calculations and comparison with single molecule thermal conductance measurements. Previous studies show that the thermal conductance of single molecules is dominated by phonons. For example, the measured thermal conductance of OPE3 is 20 ± 6 pW/K [27]. The calculated contribution from electrons and phonons are 0.1 pW/K and 19 pW/K, respectively [27]. The thermal conductance can be controlled using electrically inert side groups and phonon interference through multipath molecular backbones [17,25,26,28,29].

Table 1. Room temperature thermal conductance due to electrons (κ_e) and phonons (κ_p) of single molecules between gold electrodes using DFT–NEGF calculations and single molecule measurements.

Molecule		Calculated κ (pW/K)		Measured κ (pW/K)	Ref.
		κ_p	κ_e at DFT Fermi Energy		
Biphenyl–4,4′–dithiol (BDT)		19.6	2.3	-	[17]
2,2′–dinitro–BDT		11.7	<0.01	-	[17]
oligo(2–phenylene–4,4′–ethynylene)–dithiol (OPE2)		9.9	<0.01	-	[17]
2,2′–dinitro–OPE2		9.7	16.7	-	[17]
4,4′–bipyridyl (BP)		34.8	<0.01	-	[17]
3,3′,5,5′–tetrachloride–BP		14.8	<0.01	-	[17]
3,3′–dinitro–BP		23.6	<0.01	-	[17]
oligo(3–phenylene–4,4′–ethynylene)–dithiol (OPE3)		19	0.1	20 ± 6	[19]
Octane–dithiol		23	0.02	29 ± 8	[19]
Alkanes with dihydrobenzo[b]thiophene (BT) anchor (N = number of C2H4)	$N = 1$	25.4	0.03	-	[24]
	$N = 2$	33.4	<0.01	-	[24]
	$N = 4$	30.3	<0.01	-	[24]
	$N = 8$	5.6	<0.01	-	[24]
Alkanedithiol (N = number of C_2H_4)	$N = 1$	17–22	5.7	14.6 ± 3	[20]
	$N = 2$	18–27	1.1	13.4 ± 5	[20]
	$N = 3$	17–29	<0.01	16.9 ± 3	[20]
	$N = 4$	20–33	<0.01	26.3 ± 7	[20]
	$N = 5$	17–33	<0.01	28 ± 8	[20]
Oligoyne with BT anchor (N = number of C_2H_4)	$N = 1$	15.6	0.4	-	[24]
	$N = 2$	9.2	0.5	-	[24]
	$N = 4$	7.7	0.25	-	[24]

Table 1. *Cont.*

Molecule		Calculated κ (pW/K)		Measured κ (pW/K)	Ref.
		κ_p	κ_e at DFT Fermi Energy		
2,2′–bipyridine–BP		6	0.3	-	[18]
BP functionalized with tert–butyl nitroxide radical		2	1.45	-	[18]
C_{60} monomer		20–46.3	68–572	-	[23]
C_{60} dimer		7–7.3	0.1–1.8	-	[23]
Benzenedithiol	meta	7.5	-	-	[26]
	para	22.5	-	-	
Benzenediamine	meta	24.5	-	-	[25]
	para	25.2	-	-	
2–fluoro–1,4–diaminobenzene		24.4	2.62		[25]
2–chloro–1,4–diaminobenzene		22.2	2.7	-	[25]
2–bromo–1,4—diaminobenzene		16.9	2.8	-	[25]
2,5–dibromo–1,4–diaminobenzene		17.9	2.9	-	[25]
2,6–dibromo–1,4–diaminobenzene		10.5	2.9	-	[25]
2,3–dibromo–1,4–diaminobenzene		18	3	-	[25]
OPE3–diamine	meta	13.8	0.11	-	[25]
	para	24.5	<0.01	-	[25]

In most of the molecular junctions, molecules are contacted to the electrodes via suitable anchor groups [30]. From Table 1, the calculated thermal conductance of alkanes is different in [24] (Alkanes with dihydrobenzo[b]thiophene (BT) anchor) and [20] (Alkanedithiol). While the electrodes and the molecular backbone are the same, these calculations use different anchor groups for alkanes. Therefore, it seems that the anchor group plays a significant role in the thermal conductance of molecules. In order to understand the effect of anchor groups on thermal conductance, a systematic study of a given molecular backbone with different anchor groups is needed. For this reason, we choose OPE3 and exploited its thermal conductance with different anchor groups including pyridyl (PY), thiol (S), methyl sulphide (SMe) and carbodithioate (CS) between two gold electrodes (see Figure 1). We found that thermal conductance due to phonons is affected significantly by the choice of anchor groups. For example, thermal conductance of OPE3 decreased by a factor of 2 from CS to SMe. This is significant because the thermoelectric figure of merit (ZT) is inversely proportional to thermal conductance, and therefore ZT can be enhanced by a factor of 2 using the choice of a suitable anchor group.

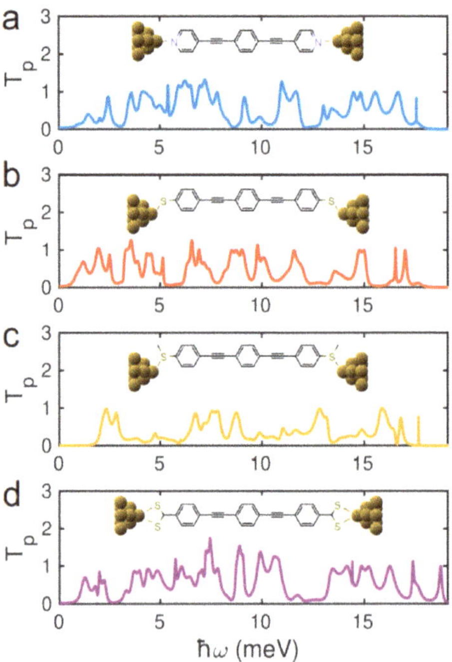

Figure 1. Phonon transmission coefficient versus phonons with energy $\hbar\omega$ for OPE3 with different anchor groups: (**a**) pyridyl PY, (**b**) thiol S, (**c**) methyl sulphide SMe and (**d**) carbodithioate CS.

2. Results and Discussion

To study vibrational and thermal properties of junctions formed by OPE3 and different anchor groups, the geometry of OPE3 in gas phase and between gold electrodes was relaxed to the force tolerance of 5 meV/Å using the SIESTA [31] implementation of density functional theory (DFT), with a double-ζ polarized basis set (DZP) and the local density approximation (LDA) functional with Ceperley and Alder (CA) parameterization. A real-space grid was defined with an equivalent energy cut-off of 350 Ry. Following the method described in [12,24], a set of xyz coordinates were generated by displacing each atom from the relaxed xyz geometry in the positive and negative x, y and z directions with $\delta q\prime = 0.01$ Å. The forces $F_i^q = \left(F_i^x, F_i^y, F_i^z \right)$ in three directions $q_i = (x_i, y_i, z_i)$ on each atom were then calculated and used to construct the dynamical matrix $D_{ij} = K_{ij}^{qq\prime}/M_{ij}$ where the mass matrix $M = \sqrt{M_i M_j}$ and $K_{ij}^{qq\prime} = \left[F_i^q\left(\delta q_j^\prime\right) - F_j^q\left(-\delta q_j^\prime\right) \right]/2\delta q_j^\prime$ for $i \neq j$ obtained from finite differences. To satisfy momentum conservation, the K for $i = j$ (diagonal terms) was calculated from $K_{ii} = -\sum_{i \neq j} K_{ij}$.

The phonon transmission $T_p(\omega)$ can then be calculated from the relation $T_p(\omega) = Trace\left(\Gamma_L^p(\omega) G_p^R(\omega) \Gamma_R^p(\omega) G_p^{R\dagger}(\omega) \right)$ where $\Gamma_{L,R}^p(\omega) = i\left(\Sigma_{L,R}^p(\omega) - \Sigma_{L,R}^p{}^\dagger(\omega) \right)$ describes the level broadening due to the coupling to the left (L) and right (R) electrodes, $\Sigma_{L,R}^p(\omega)$ is the retarded self-frequencies associated with this coupling and $G_p^R = \left(\omega^2 I - D - \Sigma_L^p - \Sigma_R^p \right)^{-1}$ is the retarded Green's function, where D and I are the dynamical and the unit matrices, respectively.

Figure 1 shows the phonon transmission coefficient T_p for phonons with energy $\hbar\omega$ traversing from one gold electrode to the other through OPE3 derivatives with different anchor groups. T_p was limited to phonons with energies $\hbar\omega < 19$ meV, which is the Debye frequency of Au electrodes [24]. The amplitude of T_p was generally higher for OPE3 with the pyridyl PY (Figure 1a) and carbodithioate CS (Figure 1d) anchors. The amplitude of T_p was noticeably lower for OPE3 with the methyl sulphide SMe anchor (Figure 1c) compared to that with the thiol S anchor (Figure 1b). This is due to a combination of two effects. First, our calculations show that the binding energy between Au-S in SMe (0.47 eV) was weaker than that of Au-S in thiol (2.08 eV). This is because SMe makes a coordination bond to Au whereas the bond between Au–S in thiol is a stronger covalent bond. This means that the vibrational coupling between Au–S is stronger with thiol compared to the SMe anchor. Secondly, phonon interference [17] due to CH_3 side group in SMe led to the suppression of T_p. The phonon waves transmitted through sulphur atoms interfered destructively with the reflected waves by CH_3 groups for given frequencies, leading to the suppression of T_p. This is like a guitar string, where waves with certain frequencies are suppressed by pressing the string at different points.

Using T_p, the phonon thermal conductance κ_p at temperature T was calculated from $\kappa_p(T) = (2\pi)^{-1} \int_0^\infty \hbar\omega T_p(\omega)(\partial f_{BE}(\omega,T)/\partial T)d\omega$ where $f_{BE}(\omega,T) = \left(e^{\hbar\omega/k_BT} - 1\right)^{-1}$ is the Bose–Einstein distribution function, $\hbar$ is the reduced Planck's constant and k_B is the Boltzmann's constant [12]. Figure 2 shows thermal conductance for OPE3 with the PY, S, SMe and CS anchor groups. κ_p increases with temperature T and saturates for temperatures higher than 150 K. This saturation of κ_p is mainly because of the small Debye frequency of gold electrodes [24]. The order of thermal conductances for different anchors is as follows: OPE3–CS (34 pW/K) > OPE3–PY (30 pW/K) > OPE3–S (25 pW/K) > OPE3–SMe (19 pW/K). Clearly, thermal conductance due to phonons is influenced strongly by the choice of anchor groups and the lowest thermal conductance is obtained for the molecule with the SMe anchor group. The thermal conductance of OPE3 with the SMe anchor was about two times lower than that with the CS anchor.

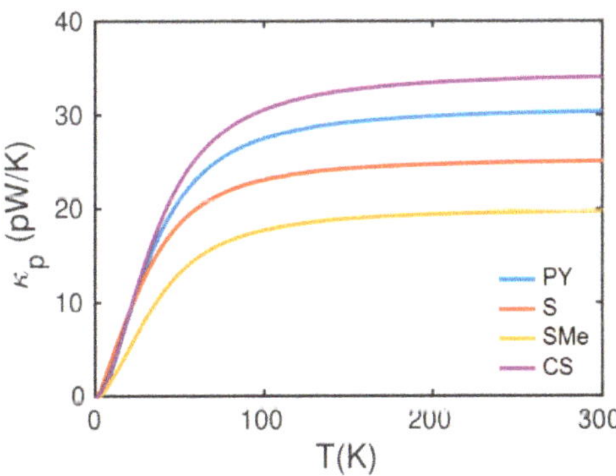

Figure 2. Phonon thermal conductance versus temperature T for OPE3 with different anchor groups.

To understand the DFT result further, we constructe a simple tight binding (TB) model of ball and springs with one degree of freedom per site and the spring constant $\gamma = 61.3 \times 10^{-3}eV$ (Figure 3a) connected to two one dimensional leads through a week coupling. Figure 3b shows the phonon transmission coefficient using the TB model for junctions with different anchor groups. Note that for simplicity, we have considered all spring constants γ the same. The phonon thermal conductance showed a similar trend to the DFT result $\kappa_p^{CS} > \kappa_p^{PY} > \kappa_p^S > \kappa_p^{SMe}$. The only difference between the junction with S and SMe anchors is the additional pendent side groups in SMe (Figure 3a). These pendent side groups attached to S clearly leads to the suppression of T_p resonances for the high frequency phonons and consequently to the decrease of the thermal conductance in OPE3 with the SMe anchor groups. The width of the T_p resonances with the CS anchor group was larger. CS anchors were connected to the electrodes from two points (inset of Figure 1d); thus the overall coupling strength to electrodes is higher. This leads to the larger broadening of T_p resonances in OPE3 with the CS anchor groups, leading to a high thermal conductance. Note that thermal conductance is proportional to the area under T_p curve that increases when width of a resonance increases.

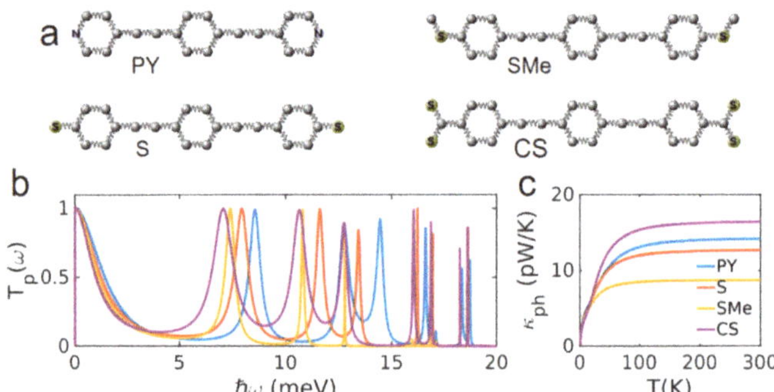

Figure 3. Tight binding (TB) model. (**a**) A simple ball and spring TB model with one degree of freedom per atom connected to two one dimensional leads through the weak coupling, (**b**) phonon transmission coefficient T_p for phonons with frequencies ω and (**c**) phonon thermal conductance versus temperature T for the simple TB model in (**a**).

The thermal conductance of OPE3 with the PY anchor is larger than that of with the S anchor. The only difference between these two junctions in the simple TB model is the additional sites at the two ends of the molecule (Figure 3a). There are two competing effects associated with this. First, the level spacing between resonances decreases when the size of the system (the number of atoms) increases. As a result, the thermal conductance is expected to increase because more resonances moved into the energy window defined by the Bose–Einstein distribution function at room temperature. Secondly, the resonance width decreases. This is because the length of the junction increases and, consequently, the density of phonon state at the two end points decreases, leading to the smaller broadening of the resonances. Clearly, the first effect is dominant here and thermal conductance increased from the S to PY anchor.

It is worth mentioning that thermal conductance is dominated by phonons in molecular junctions. For example, the room-temperature thermal conductance due to electrons and phonons in 4,4′–bipyridy connected to gold electrodes are 0.17 pW/K and 34.8 pW/K, respectively [17]. This is because the off-resonance thermal conductance due to electrons is approximately proportional to electrical conductance fromo the Wiedemann–Franz law [12,32] (e.g., $\kappa_e = \alpha G$ where G is electrical conductance and $\alpha = 7.3 \times 10^{-6}$ at room temperature). Since molecules normally show a small electrical conductance, their thermal conductance due to electrons is small. Our result demonstrates that a suitable choice of anchor group can be used to suppress thermal conductance and enhance the thermoelectric figure of merit and efficiency of molecular thermoelectric devices.

3. Conclusions

In this paper, we investigated the effect of anchor groups on thermal conductance of single OPE3 molecules. We showed that thermal conductance is affected significantly by the choice of anchor group. The thermal conductance of OPE3 can be tuned between 20–35 pW/K at room temperature by choosing different anchor groups. Our calculations indicate that SMe is the better anchor to suppress thermal conductance for thermoelectricity, whereas pyridyl and carbodithioate are better choices for thermal management applications.

Author Contributions: Conceptualization, H.S.; calculations, M.D.N., S.S. and H.S. All authors have read and agreed to the published version of the manuscript.

Funding: This research was funded by the UKRI for Future Leaders Fellowship grant number MR/S015329/2 and the Leverhulme Trust for Early Career Fellowship grant number ECF-2018-375.

Institutional Review Board Statement: Not applicable.

Informed Consent Statement: Not applicable.

Data Availability Statement: The input files to reproduce simulation data can be found at: https://warwick.ac.uk/nanolab.

Conflicts of Interest: The authors declare no conflict of interest.

References

1. Disalvo, F.J. Thermoelectric cooling and power generation. *Science* **1999**, *285*, 703–706. [CrossRef]
2. Reddy, P.; Jang, S.Y.; Segalman, R.A.; Majumdar, A. Thermoelectricity in molecular junctions. *Science* **2007**, *315*, 1568–1571. [CrossRef] [PubMed]
3. Russ, B.; Glaudell, A.; Urban, J.J.; Chabinyc, M.L.; Segalman, R.A. Organic thermoelectric materials for energy harvesting and temperature control. *Nat. Rev. Mater.* **2016**, *1*, 16050. [CrossRef]
4. Hasan, M.N.; Wahid, H.; Nayan, N.; Mohamed Ali, M.S. Inorganic thermoelectric materials: A review. *Int. J. Energy Res.* **2020**, *44*, 6170–6222. [CrossRef]
5. Evangeli, C.; Spiece, J.; Sangtarash, S.; Molina-Mendoza, A.J.; Mucientes, M., Mueller, T.; Lambert, C.; Sadeghi, H.; Kolosov, O. Nanoscale Thermal Transport in 2D Nanostructures from Cryogenic to Room Temperature. *Adv. Electron. Mater.* **2019**, *5*, 1–10. [CrossRef]
6. Chen, H.; Sangtarash, S.; Li, G.; Gantenbein, M.; Cao, W.; Alqorashi, A.; Liu, J.; Zhang, C.; Zhang, Y.; Chen, L.; et al. Exploring the thermoelectric properties of oligo(phenylene-ethynylene) derivatives. *Nanoscale* **2020**, *12*, 15150–15156. [CrossRef] [PubMed]
7. Takaloo, A.V.; Sadeghi, H. Quantum Interference Enhanced Thermoelectricity in Ferrocene Based Molecular Junctions. *J. Nanosci. Nanotechnol.* **2019**, *19*, 7452–7455. [CrossRef] [PubMed]
8. Sangtarash, S.; Sadeghi, H.; Lambert, C.J. Connectivity-driven bi-thermoelectricity in heteroatom-substituted molecular junctions. *Phys. Chem. Chem. Phys.* **2018**, *20*, 9630–9637. [CrossRef]
9. Garner, M.H.; Li, H.; Chen, Y.; Su, T.A.; Shangguan, Z.; Paley, D.W.; Liu, T.; Ng, F.; Li, H.; Xiao, S.; et al. Comprehensive suppression of single-molecule conductance using destructive σ-interference. *Nature* **2018**, *558*, 415–419. [CrossRef]
10. Zotti, L.A.; Bürkle, M.; Pauly, F.; Lee, W.; Kim, K.; Jeong, W.; Asai, Y.; Reddy, P.; Cuevas, J.C. Heat dissipation and its relation to thermopower in single-molecule junctions. *New J. Phys.* **2014**, *16*, 015004. [CrossRef]
11. Lee, W.; Kim, K.; Jeong, W.; Zotti, L.A.; Pauly, F.; Cuevas, J.C.; Reddy, P. Heat dissipation in atomic-scale junctions. *Nature* **2013**, *498*, 209–212. [CrossRef] [PubMed]
12. Sadeghi, H. Theory of electron, phonon and spin transport in nanoscale quantum devices. *Nanotechnology* **2018**, *29*, 373001. [CrossRef] [PubMed]
13. Zhao, L.-D.; Lo, S.-H.; Zhang, Y.; Sun, H.; Tan, G.; Uher, C.; Wolverton, C.; Dravid, V.P.; Kanatzidis, M.G. Ultralow thermal conductivity and high thermoelectric figure of merit in SnSe crystals. *Nature* **2014**, *508*, 373–377. [CrossRef] [PubMed]
14. Liu, J.; van der Zee, B.; Alessandri, R.; Sami, S.; Dong, J.; Nugraha, M.I.; Barker, A.J.; Rousseva, S.; Qiu, L.; Qiu, X.; et al. N-type organic thermoelectrics: Demonstration of ZT > 0.3. *Nat. Commun.* **2020**, *11*, 5694. [CrossRef] [PubMed]
15. Al-Galiby, Q.H.; Sadeghi, H.; Algharagholy, L.A.; Grace, I.; Lambert, C. Tuning the thermoelectric properties of metallo-porphyrins. *Nanoscale* **2016**, *8*, 2428–2433. [CrossRef]
16. Sangtarash, S.; Sadeghi, H.; Lambert, C.J. Exploring quantum interference in heteroatom-substituted graphene-like molecules. *Nanoscale* **2016**, *8*, 13199–13205. [CrossRef]
17. Sadeghi, H. Quantum and Phonon Interference-Enhanced Molecular-Scale Thermoelectricity. *J. Phys. Chem. C* **2019**, *123*, 12556–12562. [CrossRef]
18. Sangtarash, S.; Sadeghi, H. Radical enhancement of molecular thermoelectric efficiency. *Nanoscale Adv.* **2020**, *2*, 1031–1035. [CrossRef]
19. Mosso, N.; Sadeghi, H.; Gemma, A.; Sangtarash, S.; Drechsler, U.; Lambert, C.; Gotsmann, B. Thermal Transport through Single-Molecule Junctions. *Nano Lett.* **2019**, *19*, 7614–7622. [CrossRef]
20. Cui, L.; Hur, S.; Akbar, Z.A.; Klöckner, J.C.; Jeong, W.; Pauly, F.; Jang, S.Y.; Reddy, P.; Meyhofer, E. Thermal conductance of single-molecule junctions. *Nature* **2019**, *572*, 628–633. [CrossRef]
21. Wang, K.; Meyhofer, E.; Reddy, P. Thermal and Thermoelectric Properties of Molecular Junctions. *Adv. Funct. Mater.* **2019**, *30*, 1904534. [CrossRef]
22. Rincón-García, L.; Evangeli, C.; Rubio-Bollinger, G.; Agraït, N. Thermopower measurements in molecular junctions. *Chem. Soc. Rev.* **2016**, *45*, 4285–4306. [CrossRef]
23. Klöckner, J.C.; Siebler, R.; Cuevas, J.C.; Pauly, F. Thermal conductance and thermoelectric figure of merit of C60-based single-molecule junctions: Electrons, phonons, and photons. *Phys. Rev. B* **2017**, *95*, 245404. [CrossRef]
24. Sadeghi, H.; Sangtarash, S.; Lambert, C.J. Oligoyne Molecular Junctions for Efficient Room Temperature Thermoelectric Power Generation. *Nano Lett.* **2015**, *15*, 7467–7472. [CrossRef] [PubMed]

externally and added onto the converged system Hamiltonian after the DFT cycle. The correction is a self-energy composed of two terms: one that corrects the molecular gas-phase and one that accounts for the nonlocal polarization due the metal electrodes [25,32,33].

A typical molecular junction is composed of a single molecule bonded to electrodes on either side using chemical linker groups. Electrodes are generally metallic, and Au has been the most commonly used electrode material in scanning-probe studies because of its inertness, which enables consistent and reproducible measurements over a wide range of conditions [5,34]. Linker groups connect the molecule to the electrodes mechanically and electronically [35]. Depending on their chemical nature, linker groups bind to the electrodes either forming donor-acceptor bonds with surface asperities (for example, amine groups -NH_2) [25,36], geometry-dependent binding using pyridine groups ($-NC_5H_5$) [37,38], covalent bonding (such as thiol groups $-SH$) [39–41] or through carboxylic groups ($-COOH$) [42,43]. The atomistic details of the metal–molecule interface have been shown to tune the electronic and conducting properties of the junction [44–47]. Experimentally, variations in conductance strongly depend on the nature of the linker group. While for thiolate–Au linkers, they are very large [9,47–50], for amine-terminated molecules, these are significantly smaller [25,26,33,51]. The benzenediamine (BDA)–Au interface is a prototypical metal–molecule structure that has been extensively investigated experimentally [25,33,51] and theoretically [26,27,52,53], and is, therefore, an excellent benchmark. STM-BJ studies of BDA between Au electrodes yield a conductance peak at 6.4×10^{-3} G_0 [25,51,52].

In this work, we focus on the simulation of Au–BDA–Au junctions and investigate how sensitive the electronic and conducting properties are with respect to the atomistic termination of the Au tip structures between molecules and electrodes. We carry out this analysis using both DFT and DFT + Σ formalisms, where, for the latter, we also discuss in detail the magnitude of the necessary corrections to DFT orbital energies at the junction. This study illustrates, for a range of BDA interface geometries, how simple post-processing corrections to DFT-based transmission properties can achieve very good agreement with measured conductance values.

2. Methods

We employed the well-known SIESTA and TranSIESTA codes [13,54–56]. We constructed the junction geometries from the knowledge that the amine groups bind selectively to undercoordinated Au sites on the metal (111) surface [5,25]. We modeled the N-Au contact using three possible motifs consisting of one, three or four Au atoms (corresponding to adatom, trimer or pyramidal tips, respectively) and considered all possible combinations of these motifs. The resulting six structures are shown in Figure 1: adatom–adatom, adatom–trimer, trimer–trimer, adatom–pyramid, trimer–pyramid and pyramid–pyramid. We relaxed the positions of the atoms in the molecule and Au tip atoms until the residual Hellman–Feynman forces fell below 0.02 eV/Å. We used an exchange correlation functional which accounts for van der Waals (vdW) interactions [57]. For structure optimization, the real-space grid was defined with an equivalent energy cut-off of 250 Ry, while the reciprocal space was sampled using a $2 \times 2 \times 1$ Monkhorst-pack mesh. For calculations of density of states, we used a $5 \times 5 \times 1$ Monkhorst-pack grid.

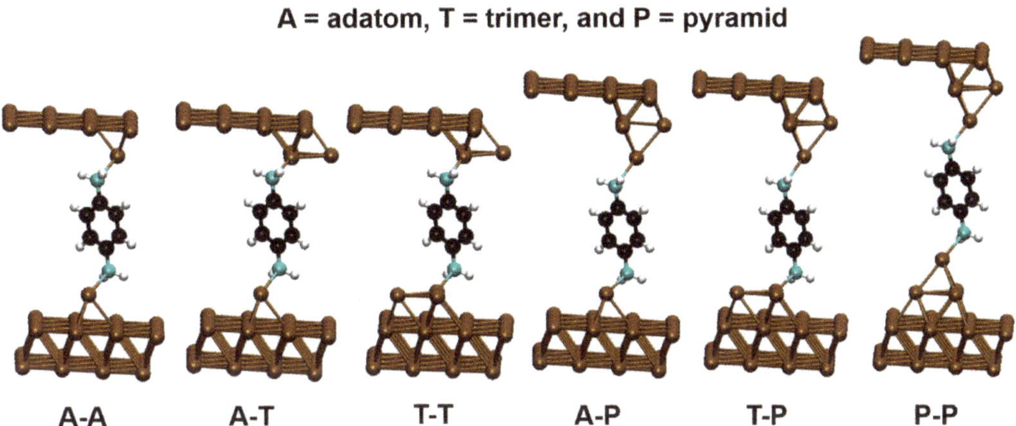

Figure 1. Model structures of benzenediamine (BDA) junctions, for combinations of different binding motifs, consisting of one, three or four Au atoms (adatom, trimer or pyramid).

Subsequent calculations of charge transport were performed at zero bias for optimized geometries using the Landauer formalism [58] as implemented by TranSIESTA [13,56]. Au atoms were described using a single-ζ polarized basis set, while a double-ζ polarized basis was used for molecular atoms. In these calculations, $5 \times 5 \times 1$ and $15 \times 15 \times 1$ Monkhorst-pack grids were used in reciprocal space for calculating the Green's function and transmission spectra, respectively.

We also employed the DFT + Σ method [25–27,29,30], which we implemented in SIESTA. In this approach, corrections are added explicitly into the system Hamiltonian (H→H + Σ) by an orbital dependent operator of the form,

$$\Sigma = \sum \Sigma_n \left| \psi_n^{mol} \; \psi_n^{mol} \right|, \tag{1}$$

where Σ_n is the self-energy correction for the nth molecular orbital, and $\left| \psi_n^{mol} \right\rangle$ denotes the wavefunction states of the molecule. These states are calculated from the Hamiltonian of the molecular subspace H_{mol}, contained into the Hamiltonian of the total system (H). The correction operator (Equation (1)) acts only on the molecular subspace $H_{mol} \subseteq H$, by construction [26,33,59,60]. The self-energy operator Σ can be constructed from a separate calculation of the relaxed isolated molecule [27,60] or from the Hamiltonian of the molecular subspace H_{mol} directly cropped from the converged ground-state Hamiltonian of the junction. Either way, the correction operator is introduced into the total Hamiltonian, which is diagonalized again in order to obtain the corrected electronic properties of the system.

The self-energy correction term consists of two parts that address: (1) the underestimated gap of the isolated molecule in conventional DFT (Σ_n^1, gas phase correction) and (2) the lack of renormalization due to the metallic electrodes (Σ_n^2, polarization due to metallic surface). The total correction is the sum of these two contributions which, as discussed below, have opposite signs, i.e., $\Sigma n = \Sigma_n^1 + \Sigma_n^2$. In principle, this correction could be calculated for every molecular orbital n. However, since we are interested in conductance, we focused on the region around the Fermi level, and calculated it only for the highest occupied molecular orbital (HOMO) and the lowest unoccupied molecular orbital (LUMO). We applied Σ_{HOMO} to all occupied states and Σ_{LUMO} to all unoccupied states [26,53].

The first contribution to the self-energy, the gas phase correction Σ_n^1, is defined as the difference between the DFT energy level and the quasiparticle energy level. We calculated the ionization potential (IP), i.e., the energy required to remove an electron from the ground state, and the electron affinity (EA), i.e., the energy required to add an electron to the ground state. These quantities are defined in terms of (DFT) total energies as IP = E(N) −

E(N − 1), and EA = E(N + 1) − E(N), where E(N$_i$) is the total energy of the system with N$_i$ electrons [28,29,61]. Furthermore, the gas-phase correction to the HOMO (LUMO) is calculated as a difference between the DFT energy position of the isolated molecule $E_{\mathrm{HOMO}}^{\mathrm{DFT}}$ ($E_{\mathrm{LUMO}}^{\mathrm{DFT}}$) and IP (EA),

$$\Sigma_{\mathrm{HOMO}}^{1} = E_{\mathrm{HOMO}}^{\mathrm{DFT}} - \mathrm{IP}, \tag{2}$$

$$\Sigma_{\mathrm{LUMO}}^{1} = -\left(E_{\mathrm{LUMO}}^{\mathrm{DFT}} - \mathrm{EA} \right) \tag{3}$$

The second term, accounting for the polarization due to the metallic surface Σ_{n}^{2}, is approximated by a classical image charge model [32,61]. It is modeled as the potential energy of a point charge distribution between two image planes. We used a point charge $q_n = 1$ for the n = HOMO or LUMO, corresponding to each atom in the molecule. Each point charge $q_{n,i}$ is located at a vertical distance z_i from the image plane z_{top} (z_{bottom}), corresponding to the top (bottom) electrode. Therefore, the self-energy term is calculated as

$$\Sigma_{n}^{2} = \sum_{i} \frac{|q_{n,i}|^{2}}{4(z_i - z_{top})} + \sum_{i} \frac{|q_{n,i}|^{2}}{4(z_i - z_{bottom})} \tag{4}$$

3. Results and Discussion

We began our analysis by studying the electronic properties of the junctions within DFT, as well as their variation with respect to the binding motifs (Figure 1). We focused only on the position of the HOMO peak, as it has been shown in the literature that HOMO dominates the zero-bias conductance for BDA [25,26,51,52]. Figure 2 shows the density of states projected over the atoms of the molecule for all the binding motifs considered: adatom–adatom, adatom–trimer, trimer–trimer, adatom–pyramid, trimer–pyramid and pyramid–pyramid. The figure is centered around the energy of the DFT HOMO peak (−1.0 eV). The inset shows the same data on an extended energy range. The data are offset vertically for clarity. From the figure, the position of the HOMO peak at the DFT level had a small variation across all different structures, ranging from −0.83 to −1.09 eV. The LUMO peak was found between 2.74 and 2.92 eV. From Figure 2, tip structures involving adatoms broadly resulted in sharper PDOS peaks and resonances closer to the Fermi level. In order to implement the DFT + Σ method, it is necessary to calculate the magnitude of the self-energy correction of the appropriate resonance (Equations (2) and (3)). We first address the calculation of the gas-phase self-energy correction of the HOMO peak, $\Sigma_{\mathrm{HOMO}}^{1}$, which involves the calculation of the isolated molecule. Several ways of computing this correction are possible. One option is to optimize the geometry of the molecule. Another possibility is to compute the molecule in the geometry it adopts at the junction. In the case of BDA, the magnitude of both corrections was the same in all cases regardless of which approach was used, −2.99 eV, although we believe that this is due to the reduced conformational flexibility of the amine linker. We anticipated a spread of values for other linkers, such as methyl-sulfide groups, which rotate when adsorbed with respect to their gas-phase geometry. In this case, it would be more consistent to use the geometry the molecule adopts at the junction [20,35].

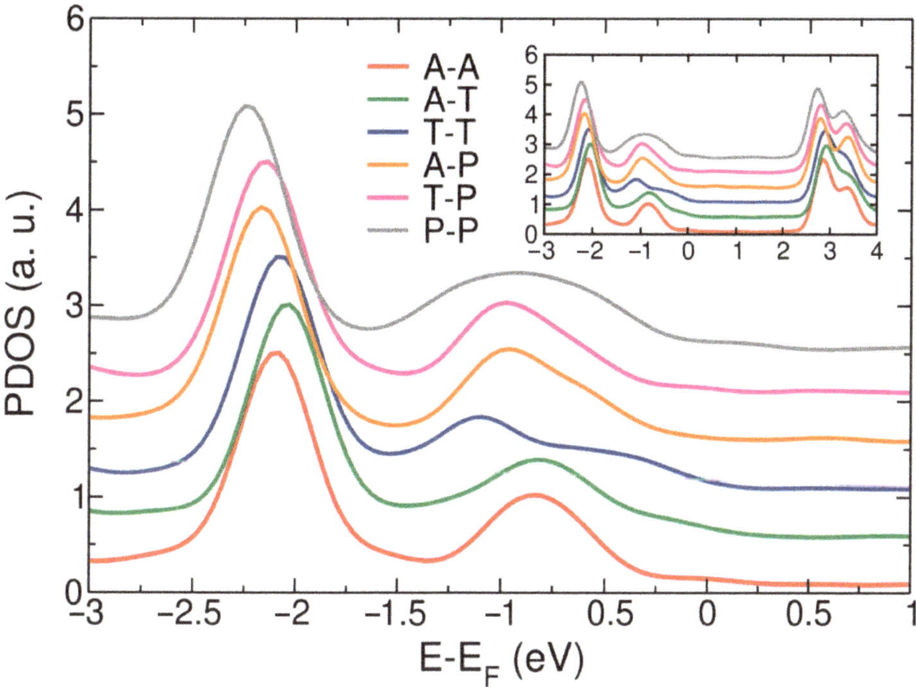

Figure 2. Density functional theory (DFT) density of states projected over molecular atoms for BDA junctions with different tip structures: adatom–adatom, adatom–trimer, trimer–trimer, adatom–pyramid, trimer–pyramid and pyramid–pyramid. Data are vertically offset for clarity. The plot is centered on the position of the HOMO peak (~−1 eV). The inset shows the same data on a wider energy range.

The second term in the self-energy correction, accounting for the polarization due to the metallic surface, Σ^2_{HOMO}, was approximated using a classical image charge model, as described in Equation (4). A single point charge was positioned at the geometrical center of the molecule [28,29,53]. The image plane position was taken to be 1.0 Å above the outer atomic plane of each Au [62]. Results for all tip structures are presented in Table 1. The calculated shift of the HOMO resonance due to screening is, for both approaches, given in Table 1. We see that the largest values were obtained for "short" tips that protrude the least from the surface, such as adatoms or trimers. When pyramidal tips were considered, image charge screening was reduced due to the larger vertical distance spanned by these tips.

Table 1. Polarization due to the metallic electrodes, Σ^2_{HOMO}, calculated using a classical image charge model with a single point charge.

	Σ^2_{HOMO} (eV)
Adatom–Adatom	1.22
Adatom–Trimer	1.22
Trimer–Trimer	1.22
Adatom–Pyramid	1.03
Trimer–Pyramid	1.03
Pyramid–Pyramid	0.81

The total self-energy correction for the HOMO peak is the sum of both (opposing) contributions. Results for all tip structures are presented in Table 2. As expected, the dependence with tip structure is opposite that of Table 1: the largest shift was found for the pyramid–pyramid combination, where the polarization self-energy was lowest, and structures with one pyramid followed. These differences are relevant for the calculation of conductance in subsequent sections. Furthermore, the total self-energy correction for the LUMO peak was 1.90 eV for the adatom–adatom, adatom–trimer and trimer–trimer tip structures, 2.09 eV for the adatom–pyramid and trimer–pyramid tip structures and 2.31 eV for the pyramid–pyramid tip structure.

Table 2. Total self-energy correction ($\Sigma_{HOMO} = \Sigma^1_{HOMO} + \Sigma^2_{HOMO}$), calculated using the geometry of the molecule at the interface and a single image point charge.

	Σ_{HOMO} (eV)
Adatom–Adatom	−1.77
Adatom–Trimer	−1.77
Trimer–Trimer	−1.77
Adatom–Pyramid	−1.96
Trimer–Pyramid	−1.96
Pyramid–Pyramid	−2.18

Having described the spread of the self-energy values with respect to tip structure, we applied the correction to the Hamiltonian ($H \rightarrow H + \Sigma$) as described in Equation (1). The shift was applied to the molecular Hamiltonian. DFT values are given by the eigenstates of the molecular box of the junction Hamiltonian, as described previously. The initial and final (corrected) energies of the HOMO resonance are given in Table 3. Figure 3 shows the corrected density of states projected onto the atoms of the molecule for tip structure combinations. The figure is centered on the range around −3.0 eV, near the position of the corrected HOMO peak. The inset reproduces the same data on a linear scale and over an extended energy range with the curves offset for clarity. The LUMO peak was shifted following the same procedure described here. The position of the corrected HOMO peak ranged between −2.9 and −3.2 eV. Additionally, the position of the corrected LUMO peak ranged between 4.5 and 4.8 eV.

Table 3. Highest occupied molecular orbital (HOMO) peak position, calculated using the DFT and DFT + Σ method.

	HOMO Peak Position (eV)	
	DFT	DFT + Σ
Adatom–Adatom	−0.84	−3.04
Adatom–Trimer	−0.83	−2.91
Trimer–Trimer	−1.09	−2.93
Adatom–Pyramid	−0.96	−3.01
Trimer–Pyramid	−0.97	−2.98
Pyramid–Pyramid	−0.93	−3.17

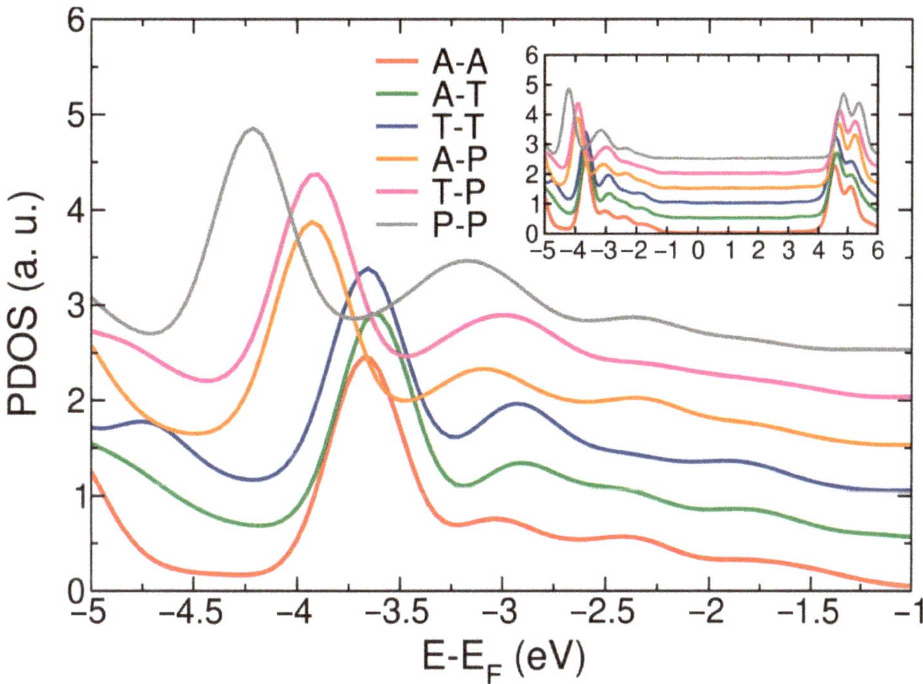

Figure 3. DFT + Σ density of states projected onto molecular atoms for BDA junctions with different tip structures. Data are offset vertically for clarity. The inset shows the same data over a wider energy range.

So far, we have focused on the DFT and corrected electronic properties, discussing how to calculate the magnitude of these corrections. In the final section of the paper, we turn to the electron transport properties and how to apply these corrections to DFT-based conductance calculations. Figure 4 shows the transmission spectra of the different BDA junctions calculated using the DFT-Landauer formalism. These calculations take the DFT-based electronic structure as input, with its well-known errors in resonance position. Figure 4 highlights the energy range below the Fermi level where the HOMO resonance, which defines zero-bias conductance, appears at the DFT level. As before, the inset plots the same data on a linear scale over an extended energy range. Figure 4 shows that low-bias conductance is determined by the tailing of the HOMO resonance into the Fermi level [25,26,51,52]. In Figure 4, the HOMO peak position is found between -0.85 and -1.00 eV over the tip structures considered, obviously following the DFT-based density of states (see Figure 2 and Table 3). This is to be compared to the HOMO transmission peak between -0.94 and -1.27 eV reported in the literature [25,26]. Calculated conductance values, given by the transmission at the Fermi level, are given in Table 4. Furthermore, as is common for DFT-based approaches, calculated conductance significantly exceeds the experimental value ($6.4 \times 10^{-3}\ G_0$) [25,51,52], in this case, by about a factor 10.

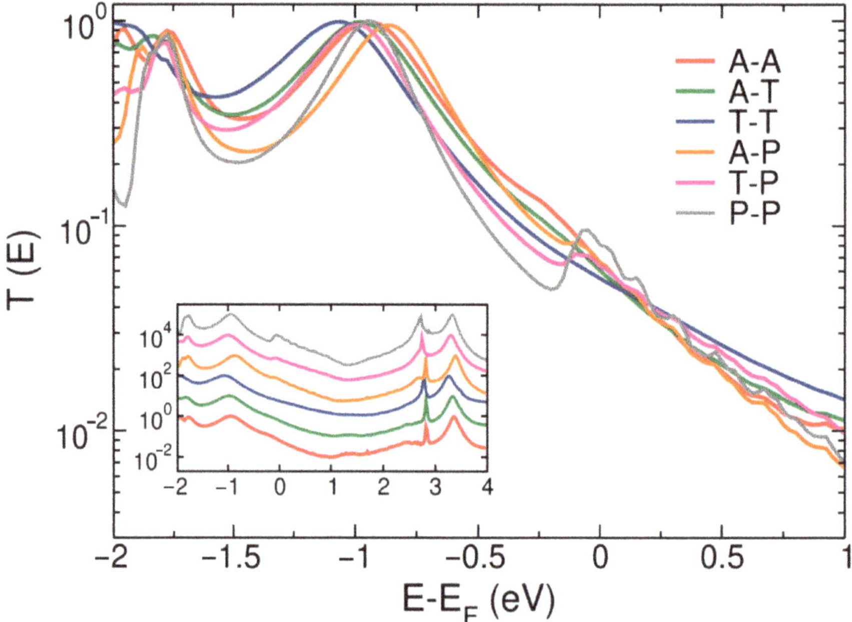

Figure 4. DFT-Landauer transmission spectra of BDA junctions using different tip structures. The figure highlights the tail of the HOMO resonance (~ -1 eV at the DFT level) into the Fermi level. The inset shows the same data over an extended energy range.

Table 4. Conductance for BDA junctions using different tip structures from DFT-Landauer and from Lorentzian fits to DFT HOMO or to DFT + Σ HOMO. Numbers in parenthesis indicate the overestimation of conductance compared to the measured value (6.4×10^{-3} G_0) [25,51,52].

	Conductance (10^{-2} G_0)		
	DFT-Landauer	Lorentzian (DFT)	Lorentzian (DFT + Σ)
Adatom–Adatom	6.61 (10.33×)	6.66 (10.40×)	1.00 (1.56×)
Adatom–Trimer	6.07 (9.48×)	6.64 (10.38×)	1.03 (1.61×)
Trimer–Trimer	5.55 (8.68×)	6.25 (9.77×)	1.08 (1.69×)
Adatom–Pyramid	6.66 (10.41×)	5.66 (8.85×)	0.64 (1.00×)
Trimer–Pyramid	6.33 (9.88×)	4.62 (7.21×)	0.63 (1.00×)
Pyramid–Pyramid	8.10 (12.65×)	3.23 (5.05×)	0.35 (1.82×)

To improve the agreement between calculated and measured conductance, it is necessary to correct the position of the conducting orbital. The implementation of DFT + Σ in a fully self-consistent DFT-NEGF cycle, out of equilibrium, is far from simple. However, trends can often be drawn from studies in equilibrium. Since in many molecular junctions, conductance takes place due to non-resonant tunneling [3,11,34], it is illustrative to consider simple models involving only one molecular resonance. Although single level models fail when transport involves several molecular orbitals [63] or in cases of quantum interference [64], they nevertheless provide a good starting point for many representative molecular junctions. Different ways of fitting the relevant parameters and their accuracy have been discussed [9,35,44,59,63–67]. Here, we consider a single level (HOMO) at zero bias.

First, the DFT-based conducting peak (in this case HOMO) is fitted using a Lorentzian model of the form,

$$T(E) = \frac{A}{\left((E - E_{\mathrm{HOMO}})^2 + (\Gamma/2)^2\right)},$$

(5)

where E_{HOMO} and A are the HOMO peak position and amplitude, respectively, and Γ is the full width at half maximum. The position of the resonance is then corrected using the self-energy previously addressed (Σ_{HOMO}). This model assumes that width of the Lorentzian (Γ) is unchanged, which is reasonable since DFT captures the electronic coupling well. For the BDA junctions considered, Γ takes three range of values. For adatom–adatom, adatom–trimer and trimer–trimer structures, the widths were 0.51, 0.52, and 0.55 eV, respectively. For adatom–pyramid and trimer–pyramid structures, the widths were 0.43 and 0.45 eV. Finally, for the pyramid–pyramid structure, the width was the lowest, 0.35 eV. This agrees well with previous calculations, where the conductance peak width showed a modest variation between 0.34 to 0.56 eV [25]. The conductance from the Lorentzian fitted curve is presented in Table 4. The agreement with DFT-Landauer values was very good, where the structures with the widest peaks showed a conductance closest to DFT-Landauer.

Table 4 also reports the conductance values calculated from the Lorentzian fit using the DFT + Σ resonance positions. Agreement with experiment was significantly improved. For the adatom–pyramid and trimer–pyramid structures, the corrected conductance was 1.0×10^{-2} G_0. This value still exceeded the experimental conductance, but only for a factor of about 1.6, a substantial improvement as compared to the previous factor of 10. For the adatom–pyramid and trimer–pyramid structures, the corrected conductance was about 6.3×10^{-3} G_0, which perfectly matches with the experimental value [25,51,52]. Furthermore, for the pyramid–pyramid structure, the corrected conductance was 3.5×10^{-3} G_0, even falling below the experimental conductance value by a factor of about 1.8. Additionally, the variation between the calculated conductance values among all tip structures was modest, about 18%. This is comparable to the experimental variation of conductance, about 8% [25]. Therefore, the combination of corrections to resonance positions within the DFT + Σ methodology, and a Lorentzian model of conductance, leads to a very good quantitative agreement with measured conductance values of BDA. The approach that we have described here would be of use to calculate the level-corrected conductance for broad classes of molecular junctions from standard DFT-Landauer calculations and rather straightforward corrections to resonance positions, combined with a Lorentzian transport model.

4. Conclusions

We studied the simulation of electronic and charge transport properties of BDA junctions with different tip structures. As is well known, at the DFT level, the HOMO position is too close to the Fermi level, and we discussed corrections to the DFT-based HOMO energy within the DFT + Σ formalism. We discussed the two contributions to the resonance energy correction, arising from the self-energy of the isolated molecule and from the screening at the metallic interface. This correction shifts the HOMO resonance further form the Fermi energy. We found that, for BDA junctions, the first term was not sensitive to the details of its calculations. However, the contribution due to interface screening, which was approximated here by a classical image charge model, did show a substantial variation of several tenths of an eV with tip structure. The total correction to the HOMO resonance at the interface was close to 2 eV towards more negative values. For the interfaces considered, DFT-based transmission spectra of BDA yielded conductance values that significantly overestimate the measured conductance by as much as an order of magnitude. We found that a Lorentzian model considering the HOMO resonance matched the DFT transmission spectra well. We showed that a DFT + Σ approach where this Lorentzian model was combined with the corrected HOMO energy produced a significant improvement in the calculated conductance values. For the different interface structures considered, this approach resulted in values in quantitative agreement with experiments.

Author Contributions: Conceptualization, H.V.; Investigation, E.M.; Methodology, E.M.; Supervision. H.V.; Writing—original draft, E.M.; Writing—review & editing, H.V. All authors have read and agreed to the published version of the manuscript.

Funding: This research was funded by the international mobility MSCA-IF II FZU grant (CZ.02.2.69/ 0.0/0.0/18_070/0010126) and the Czech Science Foundation GAČR (19-23702S). Computational resources were supplied by the project "e-Infrastruktura CZ" (e-INFRA LM2018140) provided within the program Projects of Large Research, Development and Innovations Infrastructures.

Institutional Review Board Statement: Not applicable.

Informed Consent Statement: Not applicable.

Data Availability Statement: The data presented in this study are available on reasonable request from the corresponding authors.

Conflicts of Interest: The authors declare no conflict of interest.

References

1. Aviram, A.; Ratner, M.A. Molecular rectifiers. *Chem. Phys. Lett.* **1974**, *29*, 277–283. [CrossRef]
2. Nitzan, A.; Ratner, M.A. Electron transport in molecular wire junctions. *Science* **2003**, *300*, 1384–1389. [CrossRef] [PubMed]
3. Cuevas, J.C.; Scheer, E. *Molecular Electronics: An Introduction to Theory and Experimen*; World Scientific: Singapore, 2010; Volume 1, ISBN 978-981-4282-58-1.
4. Xu, B.; Tao, N.J. Measurement of Single-Molecule Resistance by Repeated Formation of Molecular Junctions. *Science* **2003**, *301*, 1221–1223. [CrossRef] [PubMed]
5. Venkataraman, L.; Klare, J.E.; Tam, I.W.; Nuckolls, C.; Hybertsen, M.S.; Steigerwald, M.L. Single-molecule circuits with well-defined molecular conductance. *Nano Lett.* **2006**, *6*, 458–462. [CrossRef] [PubMed]
6. Tsutsui, M.; Teramae, Y.; Kurokawa, S.; Sakai, A. High-conductance states of single benzenedithiol molecules. *Appl. Phys. Lett.* **2006**, *89*, 163111. [CrossRef]
7. Martin, C.A.; Ding, D.; Van Der Zant, H.S.J.; Van Ruitenbeek, J.M. Lithographic mechanical break junctions for single-molecule measurements in vacuum: Possibilities and limitations. *New J. Phys.* **2008**, *10*, 65008. [CrossRef]
8. Haiss, W.; Wang, C.; Grace, I.; Batsanov, A.S.; Schiffrin, D.J.; Higgins, S.J.; Bryce, M.R.; Lambert, C.J.; Nichols, R.J. Precision control of single-molecule electrical junctions. *Nat. Mater.* **2006**, *5*, 995. [CrossRef]
9. Kim, Y.; Pietsch, T.; Erbe, A.; Belzig, W.; Scheer, E. Benzenedithiol: A Broad-Range Single-Channel Molecular Conductor. *Nano Lett.* **2011**, *11*, 3734–3738. [CrossRef]
10. González, M.T.; Wu, S.; Huber, R.; van der Molen, S.J.; Schönenberger, C.; Calame, M. Electrical Conductance of Molecular Junctions by a Robust Statistical Analysis. *Nano Lett.* **2006**, *6*, 2238–2242. [CrossRef]
11. Kiguchi, M. *Single-Molecule Electronics: An Introduction to Synthesis, Measurement and Theory*; Kiguchi, M., Ed.; Springer: Singapore, 2016; ISBN 9789811007248.
12. Kohn, W.; Sham, L.J. Self-Consistent Equations Including Exchange and Correlation Effects. *Phys. Rev.* **1965**, *140*, A1133–A1138. [CrossRef]
13. Brandbyge, M.; Mozos, J.L.; Ordejón, P.; Taylor, J.; Stokbro, K. Density-functional method for nonequilibrium electron transport. *Phys. Rev. B Condens. Matter Mater. Phys.* **2002**, *65*, 1654011–16540117. [CrossRef]
14. Zang, Y.; Pinkard, A.; Liu, Z.-F.; Neaton, J.B.; Steigerwald, M.L.; Roy, X.; Venkataraman, L. Electronically Transparent Au–N Bonds for Molecular Junctions. *J. Am. Chem. Soc.* **2017**, *139*, 14845–14848. [CrossRef] [PubMed]
15. Leary, E.; Zotti, L.A.; Miguel, D.; Márquez, I.R.; Palomino-Ruiz, L.; Cuerva, J.M.; Rubio-Bollinger, G.; González, M.T.; Agrait, N. The Role of Oligomeric Gold–Thiolate Units in Single-Molecule Junctions of Thiol-Anchored Molecules. *J. Phys. Chem. C* **2018**, *122*, 3211–3218. [CrossRef]
16. Li, H.; Su, T.A.; Camarasa-Gómez, M.; Hernangómez-Pérez, D.; Henn, S.E.; Pokorný, V.; Caniglia, C.D.; Inkpen, M.S.; Korytár, R.; Steigerwald, M.L.; et al. Silver Makes Better Electrical Contacts to Thiol-Terminated Silanes than Gold. *Angew. Chem. Int. Ed.* **2017**, *56*, 14145–14148. [CrossRef] [PubMed]
17. Kuang, G.; Chen, S.Z.; Yan, L.; Chen, K.Q.; Shang, X.; Liu, P.N.; Lin, N. Negative Differential Conductance in Polyporphyrin Oligomers with Nonlinear Backbones. *J. Am. Chem. Soc.* **2018**, *140*, 570–573. [CrossRef]
18. Zang, Y.; Ray, S.; Fung, E.-D.; Borges, A.; Garner, M.H.; Steigerwald, M.L.; Solomon, G.C.; Patil, S.; Venkataraman, L. Resonant Transport in Single Diketopyrrolopyrrole Junctions. *J. Am. Chem. Soc.* **2018**, *140*, 13167–13170. [CrossRef]
19. Stefani, D.; Weiland, K.J.; Skripnik, M.; Hsu, C.; Perrin, M.L.; Mayor, M.; Pauly, F.; van der Zant, H.S.J. Large Conductance Variations in a Mechanosensitive Single-Molecule Junction. *Nano Lett.* **2018**, *18*, 5981–5988. [CrossRef]
20. Schneebeli, S.; Kamenetska, M.; Foss, F.; Vazquez, H.; Skouta, R.; Hybertsen, M.; Venkataraman, L.; Breslow, R. The electrical properties of biphenylenes. *Org. Lett.* **2010**, *12*, 4114–4117. [CrossRef]
21. Perdew, J.P.; Levy, M. Physical Content of the Exact Kohn-Sham Orbital Energies: Band Gaps and Derivative Discontinuities. *Phys. Rev. Lett.* **1983**, *51*, 1884–1887. [CrossRef]

22. Kümmel, S.; Kronik, L. Orbital-dependent density functionals: Theory and applications. *Rev. Mod. Phys.* **2008**, *80*, 3–60. [CrossRef]
23. Kronik, L.; Stein, T.; Refaely-Abramson, S.; Baer, R. Excitation gaps of finite-sized systems from optimally tuned range-separated hybrid functionals. *J. Chem. Theory Comput.* **2012**, *8*, 1515–1531. [CrossRef] [PubMed]
24. Sham, L.J.; Schllter, M. Density-functional theory of the energy gap. *Phys. Rev. Lett.* **1983**, *51*, 1888–1891. [CrossRef]
25. Quek, S.Y.; Venkataraman, L.; Choi, H.J.; Louie, S.G.; Hybertsen, M.S.; Neaton, J.B. Amine—Gold linked single-molecule circuits: Experiment and theory. *Nano Lett.* **2007**, *7*, 3477–3482. [CrossRef] [PubMed]
26. Quek, S.Y.; Choi, H.J.; Louie, S.G.; Neaton, J.B. Length dependence of conductance in aromatic single-molecule junctions. *Nano Lett.* **2009**, *9*, 3949–3953. [CrossRef]
27. Li, G.; Rangel, T.; Liu, Z.F.; Cooper, V.R.; Neaton, J.B. Energy level alignment of self-assembled linear chains of benzenediamine on Au(111) from first principles. *Phys. Rev. B* **2016**, *93*, 125429. [CrossRef]
28. Markussen, T.; Jin, C.; Thygesen, K.S. Quantitatively accurate calculations of conductance and thermopower of molecular junctions. *Phys. Status Solidi Basic Res.* **2013**, *250*, 2394–2402. [CrossRef]
29. Zotti, L.A.; Bürkle, M.; Pauly, F.; Lee, W.; Kim, K.; Jeong, W.; Asai, Y.; Reddy, P.; Cuevas, J.C. Heat dissipation and its relation to thermopower in single-molecule junctions. *New J. Phys.* **2014**, *16*, 15004. [CrossRef]
30. Jin, C.; Strange, M.; Markussen, T.; Solomon, G.C.; Thygesen, K.S. Energy level alignment and quantum conductance of functionalized metal-molecule junctions: Density functional theory versus GW calculations. *J. Chem. Phys.* **2013**, *139*, 184307. [CrossRef]
31. Jin, C.; Thygesen, K.S. Dynamical image-charge effect in molecular tunnel junctions: Beyond energy level alignment. *Phys. Rev. B Condens. Matter Mater. Phys.* **2014**, *89*, 41102. [CrossRef]
32. Neaton, J.B.; Hybertsen, M.S.; Louie, S.G. Renormalization of molecular electronic levels at metal-molecule interfaces. *Phys. Rev. Lett.* **2006**, *97*, 216405. [CrossRef]
33. Dell'Angela, M.; Kladnik, G.; Cossaro, A.; Verdini, A.; Kamenetska, M.; Tamblyn, I.; Quek, S.Y.; Neaton, J.B.; Cvetko, D.; Morgante, A.; et al. Relating energy level alignment and amine-linked single molecule junction conductance. *Nano Lett.* **2010**, *10*, 2470–2474. [CrossRef] [PubMed]
34. Su, T.A.; Neupane, M.; Steigerwald, M.L.; Venkataraman, L.; Nuckolls, C. Chemical principles of single-molecule electronics. *Nat. Rev. Mater.* **2016**, *1*, 16002. [CrossRef]
35. Arasu, N.P.; Vázquez, H. Direct Au-C contacts based on biphenylene for single molecule circuits. *Phys. Chem. Chem. Phys.* **2018**, *20*, 10378–10383. [CrossRef] [PubMed]
36. Park, Y.S.; Whalley, A.C.; Kamenetska, M.; Steigerwald, M.L.; Hybertsen, M.S.; Nuckolls, C.; Venkataraman, L. Contact chemistry and single-molecule conductance: A comparison of phosphines, methyl sulfides, and amines. *J. Am. Chem. Soc.* **2007**, *129*, 15768–15769. [CrossRef]
37. Kamenetska, M.; Quek, S.Y.; Whalley, A.C.; Steigerwald, M.L.; Choi, H.J.; Louie, S.G.; Nuckolls, C.; Hybertsen, M.S.; Neaton, J.B.; Venkataraman, L. Conductance and Geometry of Pyridine-Linked Single-Molecule Junctions. *J. Am. Chem. Soc.* **2010**, *132*, 6817–6821. [CrossRef]
38. Mishchenko, A.; Zotti, L.A.; Vonlanthen, D.; Bürkle, M.; Pauly, F.; Cuevas, J.C.; Mayor, M.; Wandlowski, T. Single-Molecule Junctions Based on Nitrile-Terminated Biphenyls: A Promising New Anchoring Group. *J. Am. Chem. Soc.* **2011**, *133*, 184–187. [CrossRef]
39. Tour, J.M.; Jones, L.; Pearson, D.L.; Lamba, J.J.S.; Burgin, T.P.; Whitesides, G.M.; Allara, D.L.; Parikh, A.N.; Atre, S. Self-Assembled Monolayers and Multilayers of Conjugated Thiols, alpha, omega.-Dithiols, and Thioacetyl-Containing Adsorbates. Understanding Attachments between Potential Molecular Wires and Gold Surfaces. *J. Am. Chem. Soc.* **1995**, *117*, 9529–9534. [CrossRef]
40. González, M.T.; Leary, E.; García, R.; Verma, P.; Herranz, M.Á.; Rubio Bollinger, G.; Martín, N.; Agraït, N. Break-Junction Experiments on Acetyl-Protected Conjugated Dithiols under Different Environmental Conditions. *J. Phys. Chem. C* **2011**, *115*, 17973–17978. [CrossRef]
41. Inkpen, M.S.; Liu, Z.-F.; Li, H.; Campos, L.M.; Neaton, J.B.; Venkataraman, L. Non-chemisorbed gold–sulfur binding prevails in self-assembled monolayers. *Nat. Chem.* **2019**, *11*, 351–358. [CrossRef]
42. Paik, W.; Han, S.; Shin, W.; Kim, Y. Adsorption of Carboxylic Acids on Gold by Anodic Reaction. *Langmuir* **2003**, *19*, 4211–4216. [CrossRef]
43. Chen, F.; Li, X.; Hihath, J.; Huang, Z.; Tao, N. Effect of Anchoring Groups on Single-Molecule Conductance: Comparative Study of Thiol-, Amine-, and Carboxylic-Acid-Terminated Molecules. *J. Am. Chem. Soc.* **2006**, *128*, 15874–15881. [CrossRef] [PubMed]
44. Quek, S.Y.; Kamenetska, M.; Steigerwald, M.L.; Choi, H.J.; Louie, S.G.; Hybertsen, M.S.; Neaton, J.B.; Venkataraman, L. Mechanically controlled binary conductance switching of a single-molecule junction. *Nat. Nanotechnol.* **2009**, *4*, 230–234. [CrossRef] [PubMed]
45. Su, T.A.; Widawsky, J.R.; Li, H.; Klausen, R.S.; Leighton, J.L.; Steigerwald, M.L.; Venkataraman, L.; Nuckolls, C. Silicon ring strain creates high-conductance pathways in single-molecule circuits. *J. Am. Chem. Soc.* **2013**, *135*, 18331–18334. [CrossRef] [PubMed]
46. Kiguchi, M.; Ohto, T.; Fujii, S.; Sugiyasu, K.; Nakajima, S.; Takeuchi, M.; Nakamura, H. Single molecular resistive switch obtained via sliding multiple anchoring points and varying effective wire length. *J. Am. Chem. Soc.* **2014**, *136*, 7327–7332. [CrossRef] [PubMed]

47. Kaneko, S.; Montes, E.; Suzuki, S.; Fujii, S.; Nishino, T.; Tsukagoshi, K.; Ikeda, K.; Kano, H.; Nakamura, H.; Vázquez, H.; et al. Identifying the molecular adsorption site of a single molecule junction through combined Raman and conductance studies. *Chem. Sci.* **2019**, *10*, 6261–6269. [CrossRef] [PubMed]
48. Paulsson, M.; Krag, C.; Frederiksen, T.; Brandbyge, M. Conductance of Alkanedithiol Single-Molecule Junctions: A Molecular Dynamics Study. *Nano Lett.* **2009**, *9*, 117–121. [CrossRef]
49. Isshiki, Y.; Fujii, S.; Nishino, T.; Kiguchi, M. Fluctuation in Interface and Electronic Structure of Single-Molecule Junctions Investigated by Current versus Bias Voltage Characteristics. *J. Am. Chem. Soc.* **2018**, *140*, 3760–3767. [CrossRef]
50. Zotti, L.A.; Kirchner, T.; Cuevas, J.-C.; Pauly, F.; Huhn, T.; Scheer, E.; Erbe, A. Revealing the Role of Anchoring Groups in the Electrical Conduction Through Single-Molecule Junctions. *Small* **2010**, *6*, 1529–1535. [CrossRef]
51. Venkataraman, L.; Klare, J.E.; Nuckolls, C.; Hybertsen, M.S.; Steigerwald, M.L. Dependence of single-molecule junction conductance on molecular conformation. *Nature* **2006**, *442*, 904–907. [CrossRef]
52. Kim, T.; Vázquez, H.; Hybertsen, M.S.; Venkataraman, L. Conductance of molecular junctions formed with silver electrodes. *Nano Lett.* **2013**, *13*, 3358–3364. [CrossRef]
53. Chen, Y.; Tamblyn, I.; Quek, S.Y. Energy Level Alignment at Hybridized Organic-Metal Interfaces: The Role of Many-Electron Effects. *J. Phys. Chem. C* **2017**, *121*, 13125–13134. [CrossRef]
54. Soler, J.M.; Artacho, E.; Gale, J.D.; García, A.; Junquera, J.; Ordejón, P.; Sánchez-Portal, D. The SIESTA method for ab initio order-N materials simulation. *J. Phys. Condens. Matter* **2002**, *14*, 2745. [CrossRef]
55. García, A.; Papior, N.; Akhtar, A.; Artacho, E.; Blum, V.; Bosoni, E.; Brandimarte, P.; Brandbyge, M.; Cerdá, J.I.; Corsetti, F.; et al. Siesta: Recent developments and applications. *J. Chem. Phys.* **2020**, *152*, 204108. [CrossRef] [PubMed]
56. Papior, N.; Lorente, N.; Frederiksen, T.; García, A.; Brandbyge, M. Improvements on non-equilibrium and transport Green function techniques: The next-generation TRANSIESTA. *Comput. Phys. Commun.* **2017**, *212*, 8–24. [CrossRef]
57. Román-Pérez, G.; Soler, J.M. Efficient implementation of a van der waals density functional: Application to double-wall carbon nanotubes. *Phys. Rev. Lett.* **2009**, *103*, 96102. [CrossRef]
58. Datta, S. *Quantum Transport: Atom to Transistor*; Cambridge University Press: Cambridge, UK, 2005; ISBN 9781139164313.
59. Capozzi, B.; Chen, Q.; Darancet, P.; Kotiuga, M.; Buzzeo, M.; Neaton, J.B.; Nuckolls, C.; Venkataraman, L. Tunable Charge Transport in Single-Molecule Junctions via Electrolytic Gating. *Nano Lett.* **2014**, *14*, 1400–1404. [CrossRef]
60. Liu, Z.F.; Wei, S.; Yoon, H.; Adak, O.; Ponce, I.; Jiang, Y.; Jang, W.D.; Campos, L.M.; Venkataraman, L.; Neaton, J.B. Control of single-molecule junction conductance of porphyrins via a transition-metal center. *Nano Lett.* **2014**, *14*, 5365–5370. [CrossRef]
61. Flores, F.; Ortega, J.; Vázquez, H. Modelling energy level alignment at organic interfaces and density functional theory. *Phys. Chem. Chem. Phys.* **2009**, *11*, 8658–8675. [CrossRef]
62. Smith, N.V.; Chen, C.T.; Weinert, M. Distance of the image plane from metal surfaces. *Phys. Rev. B* **1989**, *40*, 7565–7573. [CrossRef]
63. Schneider, N.L.; Néel, N.; Andersen, N.P.; Lü, J.T.; Brandbyge, M.; Kröger, J.; Berndt, R. Spectroscopy of transmission resonances through a C60junction. *J. Phys. Condens. Matter* **2014**, *27*, 15001. [CrossRef]
64. Solomon, G.C.; Andrews, D.Q.; Hansen, T.; Goldsmith, R.H.; Wasielewski, M.R.; Van Duyne, R.P.; Ratner, M.A. Understanding quantum interference in coherent molecular conduction. *J. Chem. Phys.* **2008**, *129*, 54701. [CrossRef]
65. Fujii, S.; Marqués-González, S.; Shin, J.Y.; Shinokubo, H.; Masuda, T.; Nishino, T.; Arasu, N.P.; Vázquez, H.; Kiguchi, M. Highly-conducting molecular circuits based on antiaromaticity. *Nat. Commun.* **2017**, *8*, 15984. [CrossRef] [PubMed]
66. Refaely-Abramson, S.; Liu, Z.F.; Bruneval, F.; Neaton, J.B. First-Principles Approach to the Conductance of Covalently Bound Molecular Junctions. *J. Phys. Chem. C* **2019**, *123*, 6379–6387. [CrossRef]
67. Delmas, V.; Diez-Cabanes, V.; van Dyck, C.; Scheer, E.; Costuas, K.; Cornil, J. On the reliability of acquiring molecular junction parameters by Lorentzian fitting of I/V curves. *Phys. Chem. Chem. Phys.* **2020**, *22*, 26702–26706. [CrossRef] [PubMed]

Article

Spin Dependent Transport through Driven Magnetic System with Aubry-Andre-Harper Modulation

Arpita Koley [1], Santanu K. Maiti [1,*], Judith Helena Ojeda Silva [2,3] and David Laroze [4]

[1] Physics and Applied Mathematics Unit, Indian Statistical Institute, 203 Barrackpore Trunk Road, Kolkata 700108, India; arpitakoley94@gmail.com

[2] Grupo de Física de Materiales, Universidad Pedagógica y Tecnológica de Colombia, Tunja 150003, Colombia; judith.ojeda@uptc.edu.co

[3] Laboratorio de Química Teórica y Computacional, Grupo de Investigación Química-Física Molecular y Modelamiento Computacional (QUIMOL), Facultad de Ciencias, Universidad Pedagógica y Tecnológica de Colombia, Tunja, Boyacá 150003, Colombia

[4] Instituto de Alta Investigación, CEDENNA, Universidad de Tarapacá, Casilla 7D, Arica 1000000, Chile; dlarozen@uta.cl

* Correspondence: santanu.maiti@isical.ac.in

Abstract: In this work, we put forward a prescription of achieving spin selective electron transfer by means of light irradiation through a tight-binding (TB) magnetic chain whose site energies are modulated in the form of well known Aubry–Andre–Harper (AAH) model. The interaction of itinerant electrons with local magnetic moments in the magnetic system provides a misalignment between up and down spin channels which leads to a finite spin polarization (SP) upon locating the Fermi energy in a suitable energy zone. Both the energy channels are significantly affected by the irradiation which is directly reflected in degree of spin polarization as well as in its phase. We include the irradiation effect through Floquet ansatz and compute spin polarization coefficient by evaluating transmission probabilities using Green's function prescription. Our analysis can be utilized to investigate spin dependent transport phenomena in any driven magnetic system with quasiperiodic modulations.

Keywords: spin polarization; magnetic chain with AAH modulation; light irradiation

Citation: Koley, A.; Maiti, S.K.; Ojeda Silva, J.H.; Laroze, D. Spin Dependent Transport through Driven Magnetic System with Aubry-Andre-Harper Modulation. *Appl. Sci.* **2021**, *11*, 2309. https://doi.org/10.3390/app11052309

Academic Editors: Roberto Zivieri and Linda Angela Zotti

Received: 29 December 2020
Accepted: 1 March 2021
Published: 5 March 2021

Publisher's Note: MDPI stays neutral with regard to jurisdictional claims in published maps and institutional affiliations.

1. Introduction

"Spintronics" has been an emerging field of research during the last two decades which involves manipulation of electron spin along with its charge [1–10]. Unlike conventional electronic devices, where spin degree of freedom is neglected, spin based ones are much superior in the context of functionalities, new applications, operations and power consumption. Nowadays, spintronics can be applied almost everywhere, from data storage to robotics, speed control and navigation, designing of single as well as parallel logic gates, computer and mobile games and precise detection of defective cells, to name only a few. For most of these functionalities, spin injection across an interface is one of the important prerequisites. However, the fact is that the injection efficiency is remarkably low in most of the cases [11,12]. Though the efficiency can be improved by some mechanisms, still it is far away from the desired limit. One possible route to circumvent this issue is the use of a "spin filter" [13–20].

The basic concept to have filtration effect or in a more simple way to say to get polarized spin current from a completely unpolarized one, relies on the misalignment of up and down spin channels. That can be made possible by considering any kind of spin dependent scattering factor. One of the most common scattering mechanisms is associated with spin-orbit (SO) coupling [21–31]. In usual solid state materials SO coupling gets two different functional forms, known as Rashba [32] and Dresselhaus [33] SO couplings. The first one is involved with asymmetry in confining potential and hence can be regulated

Appl. Sci. **2021**, *11*, 2309. https://doi.org/10.3390/app11052309   https://www.mdpi.com/journal/applsci

externally, whereas the other one is associated with the bulk inversion asymmetry of the material and its strength cannot be monitored. Different SO coupled systems like molecules, semiconducting materials, tailor made systems etc., with two-, three- and even multi-terminal junction configurations have been taken into account to explore the characteristics of polarized currents [34–47]. For all these cases, especially for molecular systems the key limitation is that the SO coupling is too weak [48]. Because of this fact, large mismatch among the two spin channels is not possible which prohibits to have higher SP in a reasonable bias window.

On the other hand, due to the existence of large spin dependent scattering, magnetic materials exhibit high degree of SP compared to the SO coupled ones. Most commonly ferromagnetic (FM) materials [49,50] are used, though nowadays attention has also been paid in different kinds of antiferromagnetic (AFM) materials [51–56] for spin filtration. For efficient functioning, tuning of spin polarized current is extremely crucial and that is usually done by applying a magnetic field. However, it has several limitations especially for small size systems where confining a magnetic field is a challenging task. To eliminate this prescription some alternative proposals have been put forward by a few groups, including us. Placing the functional element within a suitable gate electrode, the degree of SP and its phase can be tuned selectively, and the gate controlled transport phenomena have also been discussed in some other contexts.

In the present work we follow a different scheme, probably not explored so far in literature, where SP is engineered by means of light irradiation [57–63]. To substantiate this fact we consider a one-dimensional (1D) FM system which is irradiated by an arbitrary polarized light (see Figure 1). Each site of the magnetic chain is associated with a local magnetic moment that interacts with the injected electron spin [64–69]. Because of this, interaction electrons get scattered. The up and down spin energy channels are largely modified due to irradiation as it renormalizes the hopping strength, and this fact is directly reflected into the transport behavior. To make the system more realistic we introduce disorder in the proposed model. Instead of "uncorrelated" disorder, we consider a "correlated" one in the form of Aubry–Andre–Harper model [70–77] since the later one exhibits several atypical signatures. Both diagonal, off-diagonal and generalized AAH systems have been extensively studied in literature exploiting several unusual phenomena, especially along the line of electronic localization, due to unique and diverse characteristic features of AAH models, and here in our present work we concentrate only on diagonal AAH system as a first attempt and discuss the interplay between the AAH potential and irradiation on spin selective electron transmission.

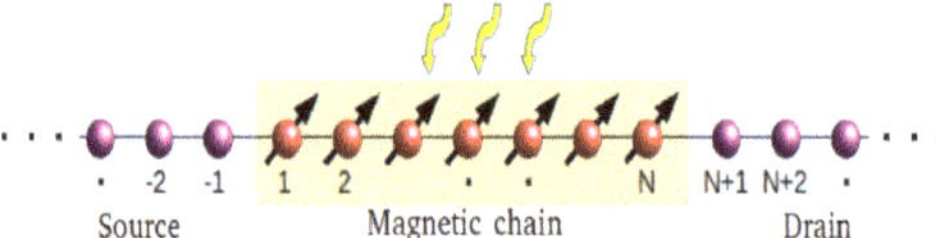

Figure 1. Spin polarized setup where a one-dimensional magnetic chain is coupled to non-magnetic source and drain electrodes. The magnetic chain is subjected to light irradiation which controls spin transfer through the junction.

Describing the quantum system within a tight-binding framework where irradiation effect is incorporated via the usual Floquet prescription [59,60], we determine spin dependent transmission probabilities following the Green's function formalism [78–82]. Using the transmission probabilities, we evaluate spin polarization coefficient. From the results, we find that the degree along with the phase of SP can be tuned in a wide range by means of irradiation. Several interesting features are emerged and our results might be useful in designing spin based electronic devices in near future.

The rest of the work is arranged as follows: In Section 2 we illustrate the spin polarized setup and TB Hamiltonian, and then give an outline of theoretical prescription for studying spin dependent transport phenomena. All the results are presented and critically analyzed in Section 3. Finally, we summarize our essential findings in Section 4.

2. Magnetic Junction, TB Hamiltonian and Theoretical Formulation

2.1. Junction Setup and the Hamiltonian

Let us begin with the spin polarized setup, shown in Figure 1, where a magnetic chain having N lattice sites (filled red balls) is clamped between two non-magnetic (NM) electrodes, commonly referred as source (S) and drain (D). Each site of the magnetic chain contains a finite magnetic moment which interacts with the injected spin σ ($\sigma = \uparrow, \downarrow$). The magnetic chain is subjected to light irradiation (yellow arrows) that plays the central role for engineering spin polarization in our analysis.

The Hamiltonian of the magnetic nanojunction can be written as

$$H = H_{mag} + H_S + H_D + H_{tun} \tag{1}$$

where different sub-Hamiltonians are associated with different parts of the junction. We describe explicitly all these terms one by one in TB framework as follows.

The sub-Hamiltonian H_{mag} reads as

$$H_{mag} = \sum_i c_i^\dagger \left(\epsilon_i - \vec{h}_i.\vec{\sigma} \right) c_i + \sum_i \left(c_{i+1}^\dagger \tilde{t} c_i + h.c. \right) \tag{2}$$

where $c_i^\dagger = \begin{pmatrix} c_{i\uparrow}^\dagger & c_{i\downarrow}^\dagger \end{pmatrix}$. $c_{i\sigma}^\dagger$ ($c_{i\sigma}$) is the usual fermionic creation (annihilation) operator. $\epsilon_i = \mathrm{diag}(\epsilon_{i\uparrow}, \epsilon_{i\downarrow})$ where $\epsilon_{i\sigma}$ represents the site energy. In the presence of AAH modulation, the site energies are expressed as [71–73] $\epsilon_{i\uparrow} = \epsilon_{i\downarrow} = W \cos(2\pi bi + \phi_v)$, where W measures the strength of the cosine modulation, b is an irrational number and ϕ_v is the AAH phase factor. In our calculations, we set $b = (1 + \sqrt{5})/2$ without loss of any generality. With a suitable setup, one can regulate the phase factor ϕ_v, and here we discuss its effect on SP. The term $\vec{h}_i.\vec{\sigma}$ is responsible for spin dependent scattering [64–69], where $\vec{h}_i$ is the spin-flip scattering factor and $\vec{\sigma}$ denotes the Pauli spin vector. It is a well known scattering phenomenon and has been elaborately studied in literature (see Refs. [64–69], and the references therein). The strength h_i is usually very large and in some cases it becomes an order of magnitude higher than the SO coupling [64]. Because of this fact, we get large mismatch between the two spin channels. The orientation of the spin flip vector is described by the conventional polar and azimuthal angles, θ_i and φ_i, respectively. Here, it is relevant to note that, in the present formulation we ignore the effect of interaction among the neighboring magnetic moments. It is well established that the moment–moment interaction can be expreesed as an effective Zeeman like term which represents the interaction of localized magnetic moments with an effective B-field (commonly referred as the "molecular field"). Compared to the interaction of itinerant electrons with local moments, as the Zeeman coupling is too weak, due to the existance of the factor μ_B, no appreciable change is expected in SP even when the magnetic field is too high.

The rest part of H_{mag} is associated with the hopping of an electron from one site to its neighboring sites. $\tilde{t} = \mathrm{diag}(\tilde{t}, \tilde{t})$. In the presence of light irradiation, the nearest-neighbor hopping (NNH) strength t gets renormalized and it takes the form [59,60]

$$\begin{aligned}
\tilde{t} &= t \frac{1}{\mathcal{T}} \int_0^{\mathcal{T}} e^{j(p-q)\Omega\tau} e^{\vec{A}.\vec{a}} \, d\tau \\
&= t J_{(p-q)}(\Lambda) \tag{3}
\end{aligned}$$

where $J_{(p-q)}$ is the $(p-q)$th order Bessel function of the first kind.

To include the effect of irradiation we follow the Floquet ansatz. Within the minimal coupling scheme the irradiation can be simplified through a vector potential $\vec{A}(\tau)$ like [59,60]

$$\vec{A}(\tau) = \left\{ A_x \sin(\Omega\tau), A_y \sin(\Omega\tau + \phi) \right\} \tag{4}$$

where A_x and A_y represent the amplitudes and ϕ corresponds to the phase. Depending on these parameters we get linear, elliptical and circularly polarized lights. The vector potential satisfies the relation $\vec{A}(\mathcal{T} + \tau) = \vec{A}(\tau)$ where $\mathcal{T}$ (=$2\pi/\Omega$) is the time period of the driving field. $\vec{a}$ is the lattice vector and $j = \sqrt{-1}$. As we are working with a strictly 1D system, Λ simplifies to $\Lambda = A_x a$, and, A_y and ϕ do not have any explicit roles in our analysis.

The other three sub-Hamiltonians of Equation (1) can be written in a much simpler way as they do not include any kind of magnetic interaction and irradiation. They are expressed as:

$$H_S = \sum_{i \leq -1} a_i^\dagger \epsilon_0 a_i + \sum_{i \leq -1} \left(a_{i+1}^\dagger t_0 a_i + h.c. \right) \tag{5}$$

$$H_D = \sum_{i \geq N+1} b_i^\dagger \epsilon_0 b_i + \sum_{i \geq N+1} \left(b_{i+1}^\dagger t_0 b_i + h.c. \right) \tag{6}$$

$$H_{tun} = c_1^\dagger t_S a_{-1} + c_N^\dagger t_D b_{N+1} + h.c. \tag{7}$$

where a_i and b_i are of the similar form like c_i, and they contain usual fermionic creation and annihilation operators. The electrodes S and D are parametrized by on-site energy ϵ_0 and t_0, respectively. The parameters t_S and t_D represent the coupling strengths of S and D with the magnetic chain, respectively.

2.2. Theoretical Formulation

The spin polarization coefficient of the above Hamiltonian (Equation (1)) is obtained by determining the spin dependent transmission probabilities which we compute by using Green's function formalism where the effects of S and D are incorporated through self-energy corrections. For comprehensive analysis of this formalism, we recommend the general readers to see the Refs. [78,79]. The effective Green's function of the magnetic chain reads as [78,79]

$$G^r = (G^a)^\dagger = \left(EI - H_{mag} - \Sigma_S - \Sigma_D \right)^{-1} \tag{8}$$

where E is the energy of the incoming electron from the source end and I is the identity matrix having dimension ($2N \times 2N$). Σ_S and Σ_D are the self-energy matrices of the S and D, respectively. From this Green's function, we determine the transmission probabilities using the Fisher–Lee expression [78–82]

$$T_{\sigma\sigma'} = \text{Tr} \left[\Gamma_S^\sigma G^r \Gamma_D^{\sigma'} G^a \right] \tag{9}$$

where Γ_S^σ and $\Gamma_D^{\sigma'}$ are the coupling matrices. These coupling matrices are found from the self-energy matrices via the relations

$$\Gamma_{S(D)}^{\sigma(\sigma')} = i \left[\Sigma_{S(D)}^{\sigma(\sigma')} - \left(\Sigma_{S(D)}^{\sigma(\sigma')} \right)^\dagger \right]. \tag{10}$$

Using Equation (9) we get both pure ($\sigma = \sigma'$) and spin flip ($\sigma \neq \sigma'$) transmissions through the magnetic junction. With these co-efficients we get the net up and down spin transmission probabilities as

$$T_\uparrow = T_{\uparrow\uparrow} + T_{\downarrow\uparrow}, \tag{11}$$
$$T_\downarrow = T_{\downarrow\downarrow} + T_{\uparrow\downarrow}. \tag{12}$$

Determining $T_\uparrow$ and $T_\downarrow$, we evaluate the spin polarization coefficient P with the relation [83–87]

$$P = \frac{T_\uparrow - T_\downarrow}{T_\uparrow + T_\downarrow}. \tag{13}$$

$P = \pm 1$ represents complete up (down) spin polarization, while $P = 0$ denotes vanishing polarization.

3. Numerical Results and Discussion

Based on the above theoretical framework now we present the results. Our primary goal is to achieve a high degree of spin polarization and tuning its phase by means of irradiation. All the results are worked out in the high frequency limit which is defined as $\hbar\Omega >> t$, and in this limiting condition only the lowest order Floquet band, i.e., $p = q = 0$ contributes. For the chosen parameter values, the frequency becomes $\Omega \geq 10^{15}$ Hz which denotes the far-infrared (FIR) region. The intensity of the irradiation is of the order of 10^7 W/m^2 which is within the experimental limit. The corresponding electric ($\mathcal{E}$) and magnetic ($B = \mathcal{E}/c$) fields are 10^5 V/m and 10^{-4} Tesla, respectively.

The other common set of parameter values that we choose for our calculations are as follows. In the electrodes S and D we take $\epsilon_0 = 0$ and $t_0 = 2$ eV, and they are coupled to the magnetic chain via the coupling strengths $t_S = t_D = 1$ eV. In the magnetic chain we choose $t = 1$ eV, AAH modulation strength $W = 1$ eV, spin flip parameter $h_i = 1$ eV $\forall i$, $\varphi_i = 0 \,\forall i$. Unless specified, we fix the total number of sites in the magnetic chain $N = 30$. The parameter values those are not constant are given in the appropriate places, and all the other energies are also measured in unit of eV. Throughout the calculations we restrict ourselves in the zero temperature limit. This is a realistic approximation as long as the average energy level spacing is higher than the thermal energy, and for small scale systems (even for $N < 200$) this condition can be easily achieved. It is also important to note that one cannot increase the chain length as much as it is possible, since we need to confine the system size within the spin coherence length. Otherwise no such phenomenon will be observed.

Let us begin with Figure 2 where spin dependent transmission probabilities T_σ along with spin polarization co-efficient P are shown as a function of energy E for some specific values of A_x. Several key features are emerged those are analyzed as follows. At a first glance we see that the transmission spectrum is highly fragmented and gapped in nature. This is solely due to the cosine modulation in site energies, as gapped energy spectrum is the generic feature of an AAH system. The transmission spectrum is a direct manifestation of the energy values. Unlike the perfect magnetic chain where energy spectrum is not gapped, for the AAH case we have a finite probability to get non-zero spin polarization at different energy zones. More importantly, even near the energy band centre we can get a reasonably large spin polarization. This feature is always desirable since one can easily place the Fermi level close to the central region, apart from placing it near the energy band edges. The role of irradiation is of course fascinating. For $A_x = 0$, a finite overlap between up and down spin transmission probabilities takes place, following the up and down spin energy channels, for a broad energy region. Therefore, for these energy zones spin polarization becomes too small. Whereas, the transmission spectra start to get shifted with A_x and they are almost separated for higher A_x, which is clearly visible by comparing the spectra given in the left column of Figure 2.

The shifting of transmission probabilities with A_x is entirely due to the modification of energy eigenvalues of the magnetic chain, since the NNH strength gets renormalized in presence of the irradiation (see Equation (3)). As effective hopping decreases compared to the irradiation free case, we get reduced allowed energy windows for up and down spin electrons, and thus the transmission spectra. All the characteristic features are directly reflected in the P-E spectra (see right column of Figure 2). Almost 100% spin polarization can be achieved for the entire allowed energy zones by suitably adjusting the irradiation parameter. It gives a clear signature of achieving externally controlled spin polarization through a magnetic nanojunction.

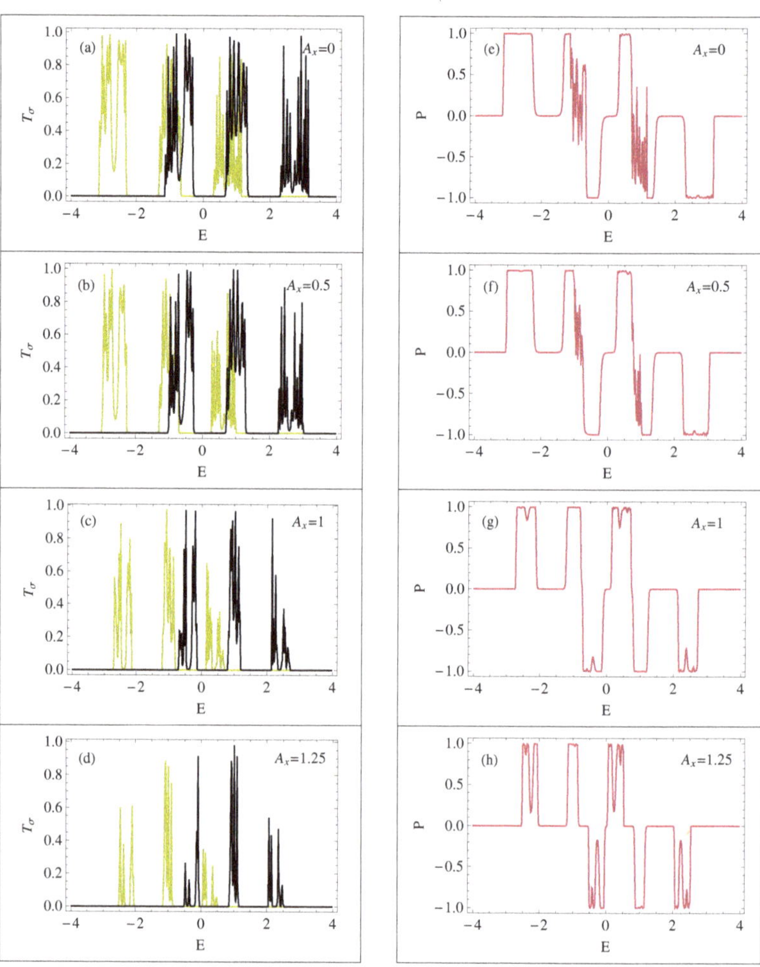

Figure 2. Up and down spin transmission probabilities ($T_\uparrow \rightarrow$ light green color and $T_\downarrow \rightarrow$ black color) and spin polarization coefficient P as a function of energy E at four typical values of A_x. Here we set $\theta_i = 0\,\forall\,i$ and AAH phase $\phi_\nu = 0$.

The above analysis gives rise to a very fundamental question that how the spin polarization can vary if we tune A_x continuously, instead of fixing it at some typical values. To demonstrate it, in Figure 3 we show the dependence of P as a function of A_x by varying it in a wide range. The results are presented for two distinct energies, $E = -1\,\text{eV}$ and $1\,\text{eV}$. Both for these two energies, P shows a large oscillation with increasing amplitude for lower A_x, and eventually saturates exhibiting 100% polarization for higher A_x. At $E = -1\,\text{eV}$ or $1\,\text{eV}$, there is a finite overlap between the two transmission functions when $A_x = 0$ and thus P becomes very small. The spectral properties and thus the transmission spectra get modified with the alteration of A_x. However, the fact is that the NNH strength t does not monotonically decrease with A_x as it follows the zeroth order Bessel function of the first kind (see Equation (3)). Due to this reason, in some cases we get finite overlap between $T_\uparrow$ and $T_\downarrow$ which yields lesser P. On the other hand when the overlap is less, higher P is achieved. This is the underlying mechanism to have oscillatory behavior of P with A_x. For large A_x, when $T_\uparrow$ and $T_\downarrow$ are practically separated we get the maximum SP. Depending on the dominating factor among $T_\uparrow$ and $T_\downarrow$, we get either 100% up or down spin polarization. These results clearly justify that the degree of SP can be tuned in a wide range by regulating A_x, without altering any other physical parameters describing the system.

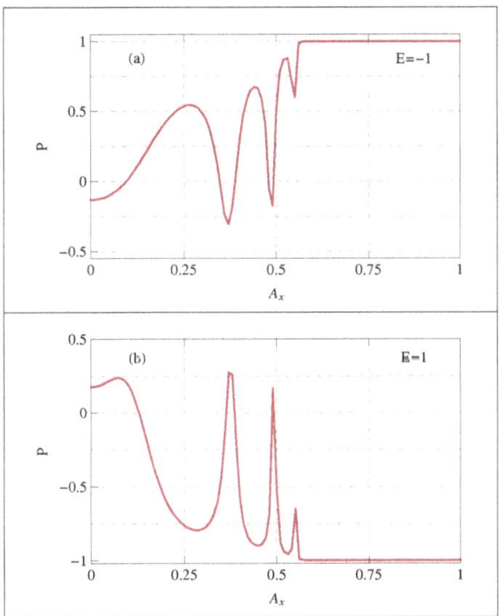

Figure 3. P-A_x characteristics at two distinct energies. All the other physical parameters kept unchanged as taken in Figure 2.

The results discussed so far are computed for the magnetic chain where all the magnetic moments are aligned along $+Z$ direction i.e., $\theta = 0$ (we refer $\theta_i = \theta$ as we assume that all the moments are aligned in a particular direction). For $\theta = 0$, there is no spin flipping i.e., $T_{\uparrow\downarrow} = T_{\downarrow\uparrow} = 0$, since in this case σ_x and σ_y do not involve into the TB Hamiltonian Equation (2) and it becomes exactly diagonal. However, finite spin flipping occurs as long as the moments are aligned in a particular direction with respect to $+Z$ axis, and to reveal the θ dependence on SP, in Figure 4, we present P-θ characteristics by changing θ in a wide range, for two different values of A_x, considering the identical energy values as taken in Figure 3. Both for the orange and black curves, associated with $A_x = 0.5$ and 1 respectively, the spin polarization co-efficient shows a complete phase reversal under rotating the magnetic moments. For the two typical values of θ, SP drops exactly to zero, as expected. With the change of θ spectral behavior gets changed and hence the SP. Thus, the alignment of the magnetic moments has an important role in SP.

To explore the explicit dependance of SP on both θ and A_x, in Figure 5, we present a density plot of P by varying these factors in a broad range fixing the energies at some specific values. From these spectra we get a clear hint about the range of physical parameters for which large degree of spin polarization can be obtained for this magnetic junction. The phase reversal is also clearly noticed.

Now we focus our attention to examine the critical role played by the AAH phase ϕ_ν on spin selective electron transfer. In Figure 6, we present the variation of spin polarization co-efficient as a function of ϕ_ν for two distinct energies. The orange curve is for $E = -0.75\,\text{eV}$, while the other one is for $E = 0.75\,\text{eV}$. A reasonably large change is reflected from both these two curves. Regulating ϕ_ν, that can be done externally, we can change the available energy channels between the electrodes as the effective site energies of the magnetic chain are modified, and therefore, the degree of spin polarization can be tuned. Thus, along with the irradiation parameter, the AAH phase can also be considered as a suitable parameter for regulating the spin transfer.

To have a more clear picture and to understand precisely the interplay between the light parameter and the AAH phase, in Figure 7, we present a density plot of SP by varying simultaneously A_x and ϕ_ν in a broad range. Some typical energies are selected like what we consider in Figure 5. Quite interestingly we find that, the degree of SP and its sign can be monitored selectively by adjusting A_x and ϕ_ν. The phenomenon persists for a wide range of these parameters which suggests that extremely fine tuning is no longer required. It certainly gives us a confidence that the present findings can be tested in a suitable laboratory setup.

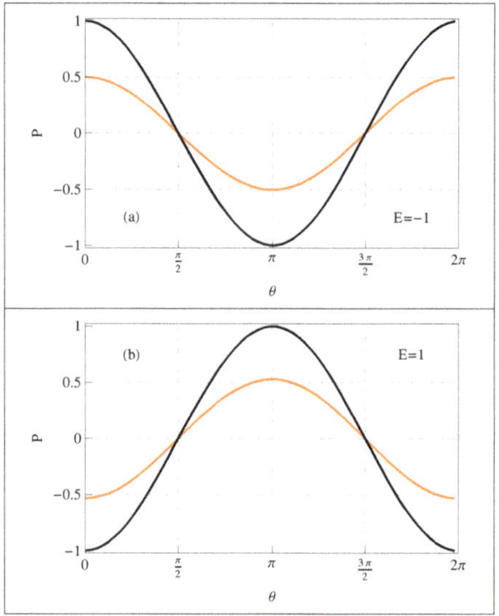

Figure 4. P as a function of θ $(\theta_i = \theta \,\forall\, i)$ at two different energies, where the orange and black lines correspond to $A_x = 0.5$ and 1, respectively. The AAH phase $\phi_\nu = 0$.

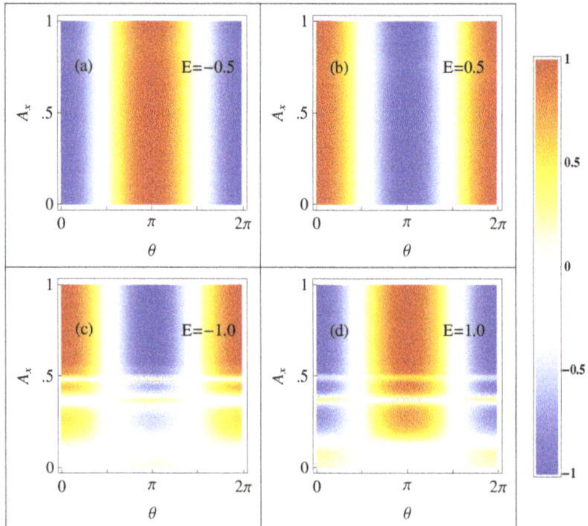

Figure 5. Simultaneous variation (density plot) of P with θ $(\theta_i = \theta \,\forall\, i)$ and A_x at some specific values of energy. The AAH phase $\phi_\nu = 0$.

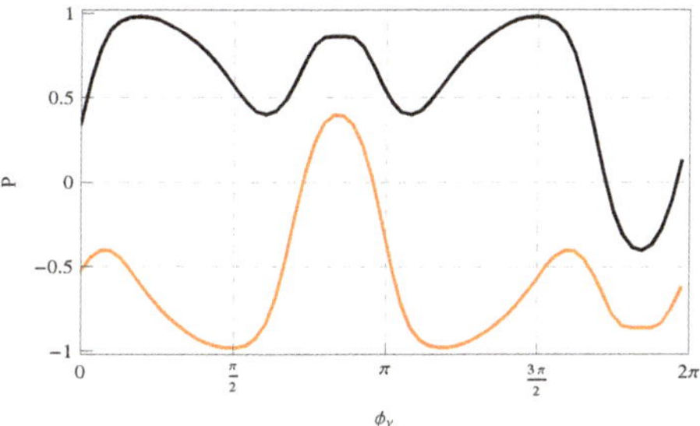

Figure 6. Spin polarization co-efficient P as a function of ϕ_v at two typical energies, where the orange and black curves are for $E = -0.75\,\mathrm{eV}$ and $E = 0.75\,\mathrm{eV}$, respectively. Here, we consider $A_x = 0.75$ and $\theta_i = 0\ \forall\, i$.

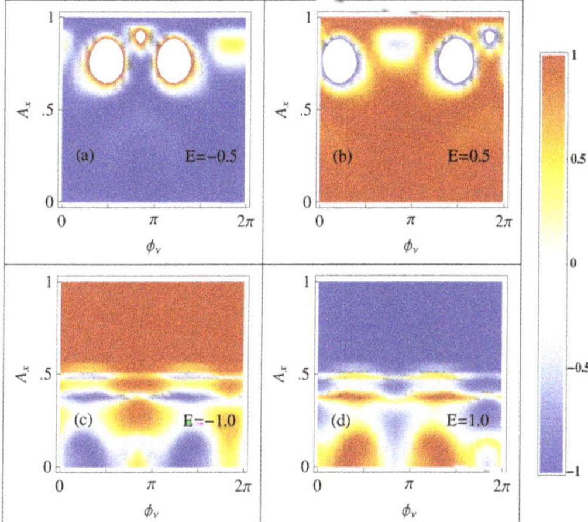

Figure 7. Simultaneous variation (density plot) of P with AAH phase ϕ_v and light parameter A_x at some specific energies. Here we set $\theta_i = 0\ \forall\, i$.

Finally, keeping in mind the possible experimental realization of our prescription it is indeed required to investigate the effect of system size on spin dependent transmission probabilities and the spin polarization co-efficient. To explore it, in Figure 8, we show the dependence of these physical quantities on system size N by varying it in a wide range at two typical energies. For both of these energies, T_σ and P exhibit large amplitude oscillation with N. This is solely due to the effect of quantum interference among electronic waves, and can be observed for other energies as well, which we confirm through our detailed numerics. The crucial thing is that the oscillating nature persists even for a reasonably large system size, and therefore, we can safely verify our proposal in a suitable laboratory setup. In this context it is relevant to note that similar kind of oscillation in transport quantities by varying system size has also been reported in different contemporary works, and for the ordered systems it can be tested even analytically [65,88,89].

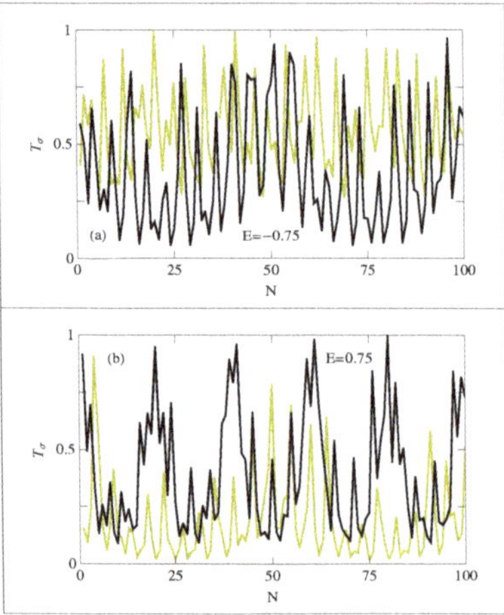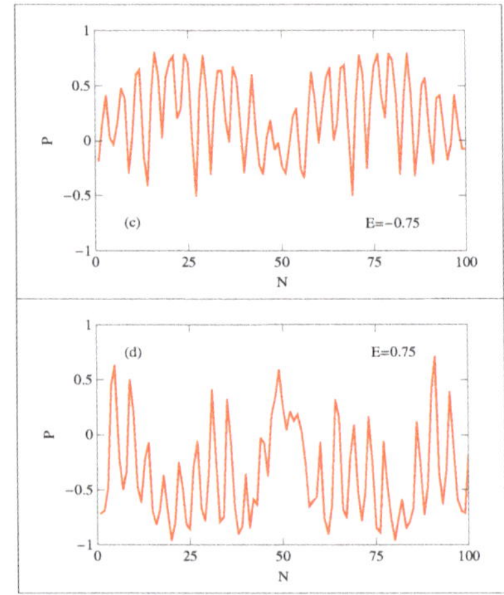

Figure 8. Spin dependent transmission probabilities ($T_\uparrow \rightarrow$ light green color and $T_\uparrow \rightarrow$ black color) and corresponding spin polarization co-efficient P as a function of system size N at two typical energies. Here we set $A_x = 0.25$, $\phi_\nu = 0$ and $\theta_i = 0 \, \forall \, i$.

4. Closing Remarks

A possible route of engineering spin polarization by means of light irradiation is proposed considering a tight-binding magnetic chain with cosine modulation in site energies. Each site of the magnetic system possesses a finite magnetic moment which interacts with the itinerent electrons. Because of this interaction, up and down spin channels get misaligned. Simulating the nanojunction, formed by placing the magnetic chain between two non-magnetic contacts, within a tight-binding framework, we determine spin dependent transmission probabilities using the Green's function formalism. From the transmission function we evaluate spin polarization co-efficient. The interplay between the irradiation, included into the Hamiltonian following the standard Floquet prescription, and the AAH potential has an important role and we investigate it critically on SP. Apart from achieving a high degree of spin polarization we can also selectively tune its phase with the help of irradiation. The peculiar gapped nature of up and down spin energy channels in presence of cosine modulation allows us to get higher filtration efficiency at multiple energy zones, and most importantly, it is also possible near the energy band centre together with other energy regions. Our analysis can be utilized to investigate spin dependent transport phenomena in different driven magnetic systems with correlated impurities.

Author Contributions: This paper was accomplished based on a collaborative work of the authors. S.K.M. conceived the project. A.K. and S.K.M. performed numerical calculations. All the authors have analyzed the data and co-wrote the paper. All authors have read and agreed to the published version of the manuscript.

Funding: This research was funded by FONDECYT 1180905 and Centers of Excellence with BASAL/ANID financing, Grant AFB180001, CEDENNA.

Institutional Review Board Statement: Not applicable.

Informed Consent Statement: Not applicable.

Data Availability Statement: The data that support the findings of this study are available from the corresponding author upon reasonable request.

Acknowledgments: AK thanks DST-SERB, Government of India (Project No. EMR/2017/000504). JHOS recognizes Universidad Pedagógica y Tecnológica de Colombia (Project No. SGI 2834).

Conflicts of Interest: The authors declare no conflict of interest.

References

1. Wolf, S.A.; Awschalom, D.D.; Buhrman, R.A.; Daughton, J.M.; von Molnár, S.; Roukes, M.L.; Chtchelkanova, A.Y.; Trege, D.M. Spintronics: A spin-based electronics vision for the future. *Science* **2001**, *294*, 1488. [CrossRef]
2. Žutić, I.; Fabian, J.; Sarma, S.D. Spintronics: Fundamentals and applications. *Rev. Mod. Phys.* **2004**, *76*, 323.
3. Sahoo, S.; Kontos, T.; Furer, J.; Hoffmann, C.; Gräber, M.; Cottet, A.; Schönenberger, C. Electric field control of spin transport. *Nat. Phys.* **2005**, *1*, 99. [CrossRef]
4. Datta, S.; Das, B. Electronic analog of the electro-optic modulator. *Appl. Phys. Lett.* **1990**, *56*, 665. [CrossRef]
5. Tsukagoshi, K.; Alphenaar, B.W.; Ago, H. Coherent transport of electron spin in a ferromagnetically contacted carbon nanotube. *Nature* **1999**, *401*, 572. [CrossRef]
6. Kobayashi, T.; Nakata, Y.; Yaji, K.; Shishidou, T.; Agterberg, D.; Yoshizawa, S.; Komori, F.; Shin, S.; Weinert, M.; Uchihashi, T.; et al. Orbital Angular Momentum Induced Spin Polarization of 2D Metallic Bands. *Phys. Rev. Lett.* **2020**, *125*, 176401. [CrossRef]
7. Yuan, L.-D.; Wang, Z.; Luo, J.-W.; Rashba, E.I.; Zunger, A. Giant momentum-dependent spin splitting in centrosymmetric low-Z antiferromagnets. *Phys. Rev. B* **2020**, *102*, 014422. [CrossRef]
8. Manchon, A.; Zelezny, J.; Miron, I.M.; Jungwirth, T.; Sinova, J.; Thiaville, A.; Garello, K.; Gambardella, P. Current-induced spin-orbit torques in ferromagnetic and antiferromagnetic systems. *Rev. Mod. Phys.* **2019**, *91*, 035004. [CrossRef]
9. Jiao, Y.; Ma, F.; Zhang, C.; Bell, J.; Sanvito, S.; Du, A. First-Principles Prediction of Spin-Polarized Multiple Dirac Rings in Manganese Fluoride. *Phys. Rev. Lett.* **2017**, *119*, 016403. [CrossRef]
10. Han, W.; Maekawa, S.; Xie, X.-C. Spin current as a probe of quantum materials. *Nat. Mat.* **2020**, *19*, 139. [CrossRef] [PubMed]
11. Prinz, G.A. Magnetoelectronics. *Science* **1998**, *282*, 1660.
12. Schmidt, G.; Ferrand, D.; Molenkamp, L.W.; Filip, A.T.; van Wees, B.J. Fundamental obstacle for electrical spin injection from a ferromagnetic metal into a diffusive semiconductor. *Phys. Rev. B* **2000**, *62*, R4790. [CrossRef]
13. Wu, M.W.; Zhou, J.; Shi, Q.W. Spin-dependent quantum transport in periodic magnetic modulations: Aharonov–Bohm ring structure as a spin filter. *Appl. Phys. Lett.* **2004**, *85*, 1012. [CrossRef]
14. Kiselev, A.A.; Kim, K.W. T-shaped ballistic spin filter. *Appl. Phys. Lett.* **2001**, *78*, 775. [CrossRef]
15. Governale, M.; Boese, D.; Zülicke, U.; Schroll, C. Filtering spin with tunnel-coupled electron wave guides. *Phys. Rev. B* **2002**, *65*, 140403. [CrossRef]
16. Maiti, S.K. Curvature effect on spin polarization in a three-terminal geometry in presence of Rashba spin-orbit interaction. *Phys. Lett. A* **2015**, *379*, 361. [CrossRef]
17. Khodas, M.; Shekhter, A.; Finkel'stein, A.M. Spin Polarization of Electrons by Nonmagnetic Heterostructures: The Basics of Spin Optics. *Phys. Rev. Lett.* **2004**, *92*, 086602. [CrossRef] [PubMed]
18. Streda, P.; Seba, P. Antisymmetric Spin Filtering in One-Dimensional Electron Systems with Uniform Spin-Orbit Coupling. *Phys. Rev. Lett.* **2003**, *90*, 256601. [CrossRef]
19. Dey, M.; Maiti, S.K.; Karmakar, S.N. Magnetic quantum wire as a spin filter: An exact study. *Phys. Lett. A* **2010**, *374*, 1522. [CrossRef]
20. Dey, M.; Maiti, S.K.; Karmakar, S.N. Spin transport through a quantum network: Effects of Rashba spin-orbit interaction and Aharonov-Bohm flux. *J. Appl. Phys.* **2011**, *109*, 024304. [CrossRef]
21. Winkler, R. *Spin-Orbit Coupling Effects in Two-Dimensional Electron and Hole Systems*; Springer Tracts in Modern Physics; Springer: New York, NY, USA, 2003; Volume 191.
22. Pareek, T.P. Pure Spin Currents and the Associated Electrical Voltage. *Phys. Rev. Lett.* **2004**, *92*, 076601. [CrossRef]
23. Sun, Q.F.; Xie, X.C. Bias-controllable intrinsic spin polarization in a quantum dot: Proposed scheme based on spin-orbit interaction. *Phys. Rev. B* **2006**, *73*, 235301. [CrossRef]
24. Maiti, S.K.; Dey, M.; Sil, S.; Chakrabarti, A.; Karmakar, S.N. Magneto-transport in a mesoscopic ring with Rashba and Dresselhaus spin-orbit interactions. *Europhys. Lett.* **2011**, *95*, 57008. [CrossRef]
25. Chi, F.; Zheng, J.; Sun, L.L. Spin-polarized current and spin accumulation in a three-terminal two quantum dots ring. *Appl. Phys. Lett.* **2008**, *92*, 172104. [CrossRef]
26. Maiti, S.K. Determination of Rashba and Dresselhaus spin-orbit fields. *J. Appl. Phys.* **2011**, *110*, 064306. [CrossRef]
27. Sil, S.; Maiti, S.K.; Chakrabarti, A. Interplay of magnetic field and geometry in magneto-transport of mesoscopic loops with Rashba and Dresselhaus spin-orbit interactions. *J. Appl. Phys.* **2012**, *112*, 024321. [CrossRef]
28. Maiti, S.K.; Sil, S.; Chakrabarti, A. A proposal for the measurement of Rashba and Dresselhaus spin-orbit interaction strengths in a single sample. *Phys. Lett. A* **2012**, *376*, 2147. [CrossRef]
29. Yin, H.-T.; Lu, T.-Q.; Liu, X.-J.; Xue, H.-J. Voltage-controllable spin-polarized transport through parallel-coupled double quantum dots with Rashba spin-orbit interaction. *Phys. Lett. A* **2009**, *373*, 285. [CrossRef]
30. Ganguly, S.; Basu, S.; Maiti, S.K. Unconventional charge and spin dependent transport properties of a graphene nanoribbon with line-disorder. *Europhys. Lett.* **2018**, *124*, 57003. [CrossRef]
31. Patra, M.; Maiti, S.K. Unconventional low-field magnetic response of a diffusive ring with spin-orbit coupling. *Phys. Lett. A* **2017**, *381*, 221. [CrossRef]
32. Bychkov, Y.A.; Rashba, E.I. Properties of a 2D electron gas with lifted spectral degeneracy. *Sov. Phys. JETP Lett.* **1984**, *39*, 78.

33. Dresselhaus, G. Spin-Orbit Coupling Effects in Zinc Blende Structures. *Phys. Rev.* **1955**, *100*, 580. [CrossRef]
34. Földi, O. Kálmán, P.; Benedict, M.G.; Peeters, F.M. Quantum rings as electron spin beam splitters. *Phys. Rev. B* **2006**, *73*, 155325. [CrossRef]
35. Hatano, N.; Shirasaki, R.; Nakamura, H. Non-Abelian gauge field theory of the spin-orbit interaction and a perfect spin filter. *Phys. Rev. A* **2007**, *75*, 032107. [CrossRef]
36. Dey, M.; Maiti, S.K.; Karmakar, S.N. Spin Hall effect in a Kagome lattice driven by Rashba spin-orbit interaction. *J. Appl. Phys.* **2012**, *112*, 024322. [CrossRef]
37. Földi, O. Kálmán, P.; Benedict, M.G.; Peeters, F.M. Networks of Quantum Nanorings: Programmable Spintronic Devices. *Nano Lett.* **2008**, *8*, 2556. [CrossRef]
38. Dey, M.; Maiti, S.K.; Sil, S.; Karmakar, S.N. Spin-orbit interaction induced spin selective transmission through a multi-terminal mesoscopic ring. *J. Appl. Phys.* **2013**, *114*, 164318. [CrossRef]
39. Cohen, G.; Hod, O.; Rabani, E. Constructing spin interference devices from nanometric rings. *Phys. Rev. B* **2007**, *76*, 235120. [CrossRef]
40. Maiti, S.K. Externally controlled selective spin transfer through a two-terminal bridge setup. *Eur. Phys. J. B* **2015**, *88*, 172. [CrossRef]
41. Földi, O.; Kálmán, P.; Peeters, F.M. Stability of spintronic devices based on quantum ring networks. *Phys. Rev. B* **2009**, *80*, 125324. [CrossRef]
42. Patra, M.; Maiti, S.K. Externally controlled high degree of spin polarization and spin inversion in a conducting junction: Two new approaches. *Sci. Rep.* **2017**, *7*, 14313. [CrossRef]
43. Göhler, B.; Hamelbeck, V.; Markus, T.Z.; Kettner, M.; Hanne, G.F.; Vager, Z.; Naaman, R.; Zacharias, H. Spin Selectivity in Electron Transmission Through Self-Assembled Monolayers of Double-Stranded DNA. *Science* **2011**, *331*, 894. [CrossRef] [PubMed]
44. Ganguly, S.; Basu, S.; Maiti, S.K. Interface sensitivity on spin transport through a three-terminal graphene nanoribbon. *Superlattices Microstruct.* **2018**, *120*, 650. [CrossRef]
45. Guo, A.-M.; Sun, Q.-F. Spin-Selective Transport of Electrons in DNA Double Helix. *Phys. Rev. Lett.* **2012**, *108*, 218102. [CrossRef]
46. Ganguly, S.; Basu, S.; Maiti, S.K. Controlled engineering of spin polarized transport properties in a zigzag graphene nanojunction. *Europhys. Lett.* **2018**, *124*, 17005. [CrossRef]
47. Guo, A.-M.; Sun, Q.-F. Spin-dependent electron transport in protein-like single-helical molecules. *Proc. Natl. Acad. Sci. USA* **2014**, *111*, 11658. [CrossRef] [PubMed]
48. Kuemmeth, F.; Ilani, S.; Ralph, D.C.; McEuen, P.L. Coupling of spin and orbital motion of electrons in carbon nanotubes. *Nature* **2008**, *452*, 448. [CrossRef] [PubMed]
49. Long, W.; Sun, Q.F.; Guo, H.; Wang, J. Gate-controllable spin battery. *Appl. Phys. Lett.* **2003**, *83*, 1397. [CrossRef]
50. Zhang, P.; Xue, Q.K.; Xie, X.C. Spin Current through a Quantum Dot in the Presence of an Oscillating Magnetic Field. *Phys. Rev. Lett.* **2003**, *91*, 196602. [CrossRef]
51. Adachi, H.; Ino, H. A ferromagnet having no net magnetic moment. *Nature* **1999**, *401*, 148. [CrossRef]
52. Xu, Y.; Wang, S.; Xia, K. Spin-Transfer Torques in Antiferromagnetic Metals from First Principles. *Phys. Rev. Lett.* **2008**, *100*, 226602. [CrossRef]
53. Jungwirth, T.; Sinova, J.; Matri, X.; Wunderlich, J.; Felser, C. The multiple directions of antiferromagnetic spintronics. *Nat. Phys.* **2018**, *14*, 200. [CrossRef]
54. Jungfleisch, M.B.; Zhang, W.; Hoffmann, A. Perspectives of antiferromagnetic spintronics. *Phys. Lett. A* **2018**, *13*, 382. [CrossRef]
55. Duine, R.A.; Lee, K.; Parkin, S.P.; Stiles, M.D. Synthetic antiferromagnetic spintronics. *Nat. Phys.* **2018**, *14*, 217. [CrossRef] [PubMed]
56. Gupta, D.D.; Maiti, S.K. Can a sample having zero net magnetization produce polarized spin current? *J. Phys. Condens. Matter* **2020**, *32*, 505803. [CrossRef]
57. Grifoni, M.; Hänggi, P. Driven quantum tunneling. *Phys. Rep.* **1998**, *304*, 229. [CrossRef]
58. Kohler, S.; Lehmann, J.; Hänggi, P. Driven quantum transport on the nanoscale. *Phys. Rep.* **2005**, *406*, 379. [CrossRef]
59. Delplace, P.; Gómez-Lexoxn, A.; Platero, G. Merging of Dirac points and Floquet topological transitions in ac-driven graphene. *Phys. Rev. B* **2013**, *88*, 245422. [CrossRef]
60. Gómez-Lexoxn, A.; Platero, G. Floquet-Bloch Theory and Topology in Periodically Driven Lattices. *Phys. Rev. Lett.* **2013**, *110*, 200403. [CrossRef] [PubMed]
61. Eckardt, A.; Anisimovas, E. High-frequency approximation for periodically driven quantum systems from a Floquet-space perspective. *New. J. Phys.* **2015**, *17*, 093039. [CrossRef]
62. Goldman, N.; Dalibard, J.; Aidelsburger, M.; Cooper, N.R. Periodically driven quantum matter: The case of resonant modulations. *Phys. Rev. A* **2015**, *91*, 033632. [CrossRef]
63. Sarkar, M.; Dey, M.; Maiti, S.K.; Sil, S. Engineering spin polarization in a driven multistranded magnetic quantum network. *Phys. Rev. B* **2020**, *102*, 195435. [CrossRef]
64. Su, Y.-H.; Chen, S.-H.; Hu, C.D.; Chang, C.-R. Competition between spin-orbit interaction and exchange coupling within a honeycomb lattice ribbon. *J. Phys. D Appl. Phys.* **2016**, *49*, 015305. [CrossRef]
65. Sarkar, S.; Maiti, S.K. Spin-selective transmission through a single-stranded magnetic helix. *Phys. Rev. B* **2019**, *100*, 205402. [CrossRef]

66. Shokri, A.A.; Mardaani, M.; Esfarjani, K. Spin filtering and spin diode devices in quantum wire systems. *Physica E* **2005**, *27*, 325. [CrossRef]
67. Shokri, A.A.; Mardaani, M. Spin-flip effect on electrical transport in magnetic quantum wire systems. *Solid State Commun.* **2006**, *137*, 53. [CrossRef]
68. Peters, R.; Kawakami, N. Ferromagnetic state in the one-dimensional Kondo lattice model. *Phys. Rev. B* **2012**, *86*, 165107. [CrossRef]
69. Peters, R.; Kawakami, N.; Pruschke, T. Spin-Selective Kondo Insulator: Cooperation of Ferromagnetism and the Kondo Effect. *Phys. Rev. Lett.* **2012**, *108*, 086402. [CrossRef]
70. Aubry, S.; André, G. Analyticity breaking and Anderson localization in incommensurate lattices. *Ann. Isr. Phys. Soc.* **1980**, *3*, 133.
71. Kraus, Y.E.; Lahini, Y.; Ringel, Z.; Verbin, M.; Zilberberg, O. Topological States and Adiabatic Pumping in Quasicrystals. *Phys. Rev. Lett.* **2012**, *109*, 106402. [CrossRef] [PubMed]
72. Rossignolo, M.; Dell'Anna, L. Localization transitions and mobility edges in coupled Aubry-André chains. *Phys. Rev. B* **2019**, *99*, 054211. [CrossRef]
73. Sil, S.; Maiti, S.K.; Chakrabarti, A. Metal-insulator transition in an aperiodic ladder network: An exact result. *Phys. Rev. Lett.* **2008**, *101*, 076803. [CrossRef] [PubMed]
74. Roy, S.; Maiti, S.K. Tight-binding quantum network with cosine modulations: Electronic localization and delocalization. *Eur. Phys. J. B* **2019**, *92*, 267. [CrossRef]
75. Saha, S.; Maiti, S.K.; Karmakar, S.N. Multiple mobility edges in a 1D Aubry chain with Hubbard interaction in presence of electric field: Controlled electron transport. *Physica E* **2016**, *83*, 358. [CrossRef]
76. Maiti, S.K.; Sil, S.; Chakrabarti, A. Phase controlled metal-insulator transition in multi-leg quasiperiodic optical lattices. *Ann. Phys. (N. Y.)* **2017**, *382*, 150. [CrossRef]
77. Patra, M.; Maiti, S.K. Controlled charge and spin current rectifications in a spin polarized device. *J. Magn. Magn. Mater.* **2019**, *484*, 408. [CrossRef]
78. Datta, S. *Electronic Transport in Mesoscopic Systems*; Cambridge University Press: Cambridge, UK, 1997.
79. Datta, S. *Quantum Transport: Atom to Transistor*; Cambridge University Press: Cambridge, UK, 2005.
80. Fisher, D.S.; Lee, P.A. Relation between conductivity and transmission matrix. *Phys. Rev. B* **1981**, *23*, 6851. [CrossRef]
81. Ventra, M.D. *Electrical Transport in Nanoscale Systems*; Cambridge University Press: Cambridge, UK, 2008.
82. Nikolić, B.K.; Allen, P.B. Quantum transport in ballistic conductors: Evolution from conductance quantization to resonant tunnelling. *J. Phys. Condens. Matter* **2000**, *12*, 9629. [CrossRef]
83. Prinz, G.A. Spin-Polarized Transport. *Phys. Today* **1995**, *48*, 58. [CrossRef]
84. Jedema, F.F.; Filip, A.T.; van Wees, B.J. Electrical spin injection and accumulation at room temperature in an all-metal mesoscopic spin valve. *Nature* **2001**, *410*, 345. [CrossRef]
85. Maekawa, S.; Shinjo, T. (Eds.) *Spin Dependent Transport in Magnetic Nanostructures*; Taylor & Francis: London, UK, 2002.
86. Nikolić, B.K.; Souma, S. Decoherence of transported spin in multichannel spin-orbit-coupled spintronic devices: Scattering approach to spin-density matrix from the ballistic to the localized regime. *Phys. Rev. B* **2005**, *71*, 195328. [CrossRef]
87. Farokhnezhad, M.; Esmaeilzadeh, M.; Ahmadi, S.; Pournaghavi, N. Controllable spin polarization and spin filtering in a zigzag silicene nanoribbon. *J. Appl. Phys.* **2015**, *117*, 173913. [CrossRef]
88. Sun, Q.-F.; Guo, A.-M. Enhanced spin-polarized transport through DNA double helix by gate voltage. *Phys. Rev. B* **2012**, *86*, 035424.
89. Martinez, D.F.; Molina, R.A.; Hu, B. Length-dependent oscillations in the dc conductance of laser-driven quantum wires. *Phys. Rev. B* **2008**, *78*, 045428. [CrossRef]

Review

Nanofabrication Techniques in Large-Area Molecular Electronic Devices

Lucía Herrer, Santiago Martín and Pilar Cea *

Departamento de Química Física, Facultad de Ciencias, Universidad de Zaragoza, Pedro Cerbuna 12, 50009 Zaragoza, Spain; lucia.h@unizar.es (L.H.); smartins@unizar.es (S.M.)
* Correspondence: pilarcea@unizar.es

Received: 30 July 2020; Accepted: 27 August 2020; Published: 1 September 2020

Abstract: The societal impact of the electronics industry is enormous—not to mention how this industry impinges on the global economy. The foreseen limits of the current technology—technical, economic, and sustainability issues—open the door to the search for successor technologies. In this context, molecular electronics has emerged as a promising candidate that, at least in the short-term, will not likely replace our silicon-based electronics, but improve its performance through a nascent hybrid technology. Such technology will take advantage of both the small dimensions of the molecules and new functionalities resulting from the quantum effects that govern the properties at the molecular scale. An optimization of interface engineering and integration of molecules to form densely integrated individually addressable arrays of molecules are two crucial aspects in the molecular electronics field. These challenges should be met to establish the bridge between organic functional materials and hard electronics required for the incorporation of such hybrid technology in the market. In this review, the most advanced methods for fabricating large-area molecular electronic devices are presented, highlighting their advantages and limitations. Special emphasis is focused on bottom-up methodologies for the fabrication of well-ordered and tightly-packed monolayers onto the bottom electrode, followed by a description of the top-contact deposition methods so far used.

Keywords: molecular electronics; self-assembly films; Langmuir-Blodgett films; electrografting; top-contact electrode

---

1. Introduction

The impact of the omnipresent Complementary Metal-Oxide Semiconductor (CMOS) electronic industry on the global economy is enormous, due to its role not only in the production of computers, mobiles, tablets, etc., but mainly as an essential component in products from many other industries (automobile, aeronautics, artificial satellites, trains, security and armies, communication systems, computer science, robotics, energy, financial services, diagnostic equipment in hospitals, booster of the research, development, and innovation, etc.). Thus, the global market for electronic components is expected to grow in the 2020–2025 period at a compound annual growth rate of ca. 4.8% [1]. The new scenario imposed by the COVID-19 pandemic has further evidenced the relevance of the electronic industry in our society. Could the reader imagine how the months of confinement would have been like without the internet, teleconferencing, online teaching, streaming videos, online shopping, etc.?

If we look back over the historical development of the electronic industry, the last four decades have witnessed enormous and rapid progress in the miniaturization of electronic devices (Moore law) from 3 µm transistors in 1980 to the current 7 nm technology already in the market, using the FINFET (Fin Field-Effect Transistor) technology, with microprocessors that incorporate more than 50 billion transistors and pursuing the 5 nm transistor in 2020. This miniaturization is accompanied by a reduction in the global size and weight of electronic components, an increase in switching performance and faster processors, an increase in logic area efficiency, as well as a decrease in the energy consumption and consequently longer battery life. It is expected that if the surface area of a transistor decreases by two every two years (reduction of the lateral size by 1.4), the 3 nm technology could appear in 2022 (the MIT has created a 2.5 nm transistor [2]), and by 2024, 1 nm transistors might emerge. There are, however, a number of associated technological problems within this miniaturization race that will need to be overcome for these predictions to be reached. In devices thinned to a few nanometers, quantum effects governing the electron behavior appear, which can make transistors unreliable, due to the quantum uncertainties [3]. The reduction in the size of transistors is also associated with a dramatic increase in the fabrication process costs (second Moore law) which has resulted in a drastic decrease of foundries (from 20 foundries for 130 nm technology to only four major companies providing transistors in the 10 to 7 nm range), which is ultimately reflected in the cost of electronic devices in the market and could make future nodes unaffordable, due to the manufacturing costs for ultra-large-scale implementation [4]. Additionally, when an increasing number of transistors are ensemble into a small area of a single piece of an integrated circuit, inelastic scattering of electrons [5] results in waste-heat that emerges as an additional problem as chips get too hot, which requires efficient heat dissipation systems [3].

The technological and economic limits of the current CMOS technology are therefore imminent, and scientists all over the world are working on alternative technologies, due to the continual and growing social demand for more efficient, more rapid, more versatile, and low-power devices, not to mention flexible electronics. In 2016, the eighteen years old ITRS [6] (International Technology Roadmap for Semiconductors) was renamed as IRDS [7] (International Roadmap for Devices and Systems), which is a clear reflection of the need of alternative technologies in the XXI century. A potential successor of our current technology, must not only fulfill the same expectations as CMOS but also outperform prevalent technology in at least a number of several clue aspects, including power consumption, mass production, fabrication costs, and performance, and overcome current functionalities, i.e., fully enter in the "More than Moore" path [8]. These efforts are being made from a multidisciplinary point of view, for which the contribution of physicists, chemists, and engineers is essential for overcoming the enormous challenges ahead.

In the above-described scenario, molecular electronics emerges as a great promise. Molecular electronics is based on the idea of using molecules as functional units in circuitry to permit, control, and manipulate the movement of electrical charges between two electrodes [9]. The fundamental tool for understanding electrical transport through these two electrodes is the creation of electrode | molecule | electrode molecular junctions [10–14]. The official birth of molecular electronics is widely recognized as 1974, with the publication of the seminal paper from Aviran and Ratner that proposed (theoretically) that a single-molecule could act as a rectifier [15]. Intense work in the field for more than four decades has included the development of methodologies based on scanning tunneling microscopy (STM) or conducting atomic force microscopy (c-AFM) for measuring the electrical properties of single-molecules and molecular assemblies [14,16–22]. These studies have resulted in a growing understanding of the key parameters that determine the electrical properties of molecular junctions (molecular backbone [12,23], chemical anchoring groups [24–26], conformation [27], metal complexation [28], redox state [18,29–31], electrode material [32–35], and if applicable, the characteristics of the medium: Solvent [36], pH [37], etc.), as well as the mechanisms behind electronic transport in molecular junctions [38–40]. In particular, molecular wires [41], switches [42], diodes [43], rectifiers [44], transistors [45], and single-molecule light-emitting diodes [46] have already been demonstrated in the laboratory and make feasible the idea of integrating molecules into electrical circuits. The expectations in molecular electronics lie on its several advantages as compared to the above-described Si-based technology with relevant contributions in the field [47–50], making emphasis on applications of molecular electronics in solar-energy harvesting, thermoelectricity, catalysis, or molecular sensing. Another field of remarkable importance today is the study of single-molecule chemical reactions within the molecular junction [51–53]. The use of molecules has several advantages—both from a fundamental and an applied point of view. The most remarkable phenomena observed in molecular junctions and their promising perspectives in a short-term future technology result from:

- The size of the molecules—in the order of a few nanometers—that may enable heightened capacities, faster performance, and high integration density, with millions of identical electric machines each of them as small as one molecule [50].

- Molecular junctions exhibit a plethora of rich and tunable physicochemical properties—different from those exhibited by molecular materials in bulk—since charge transport is governed by quantum mechanics at the molecular scale. These phenomena at the nanoscale include quantum mechanical interference, the Coulomb blockade, and the Kondo effect [12,54].

- The versatility of organic chemistry to produce millions of identical functional units (molecules)—estimated in 10^{60} molecules with 15 atoms or fewer [51]—is expected to result in not only in low-cost manufacturing, due to self-assembly (SA) capabilities, but also in thinner and low-weight devices, lower supply-voltages [51], as well a large variety of new and distinctive functionalities provided by molecules (optical, magnetic, thermoelectric, electromechanical, etc.), which are often not possible by employing conventional materials [10].

- Redox-active molecular components can be addressed and 'switched' through the introduction of a third 'gate' electrode [45,55–57]. In this field, exploration of the molecular structure, ligands, inclusion of different metal clusters, control of electrochemical gating together with the potential for modular construction [58] may have relevant implications in the field of molecular electronics.

- The properties of (magnetic) molecules together with control over the spin state of molecular devices by tuning the interaction of the localized orbitals of the molecule with the electronic states of the electrode also open the door to the exploration and control of spin transport phenomena and spintronic applications, e.g., switches and qubits [59–62]. In addition, the study of magnetic spins on a molecule connected to a superconducting electrode represents a fascinating topic of interest today [63].

- One of the most promised envisioned applications of molecular electronics is thermoelectrics, i.e., efficient conversion of heat to electricity [64]. Fabrication of ultrahigh-efficient thermoelectric power generators as small as one molecule is possible, and could potentially be used to build ultrahigh-efficient thermoelectric power generators [64–66]. These power converters could reuse and recycle the dissipation heat produced by (molecular) electronic devices into electricity to (partially) supply the power required to operate the device, resulting in a meaningful decrease in electric power consumption [67], and also in applications for on-chip cooling in nanoscale electronic devices [68].

- The combined use of organic molecules and carbon-based or polymeric-based materials as electrodes could pave the road towards not only flexible devices [33,69], but also to biocompatible electronics [70] and also lead to more sustainable fabrication processes, reduce e-waste and culminate in the development of electronics that self-degrade after service life [71].

Two different paradigms have attracted the interest of researchers in the field of molecular electronics. On the one hand, the study of single-molecule junctions, i.e., electrode | molecule | electrode, is of fundamental importance for the understanding of charge transport, as well as the different factors that determine the electrical properties of these junctions, Figure 1a. Thus, it is well known that the molecular structure, the anchoring groups, the presence of metal or redox moieties, etc. can strongly determine charge transport at the molecular level. However, fabrication of these single molecular junctions with our current technology is not a scalable process, which seriously hampers the incorporation of molecular electronics into the market. On the other hand, the study of large-area devices, i.e., electrode | molecular assembly | electrode, has gained attention in the last years. Here, collective effects (intermolecular forces and polarization phenomena), induced by the close packing of the molecules in the monolayer, may determine the electrical behavior of these systems, which opens the door to finding new effects and also to tune the properties of the devices by an exhaustive control of interactions between neighbor molecules, Figure 1b. Additionally, these molecular assemblies allow the manufacture of devices with a surface density of up to 10^{15} molecules·cm^{-2}. Importantly, the deposition of molecules onto the electrode can be carried out using simple and scalable technologies that allow high-quality and reproducible assemblies, making possible the manufacture of thousands of devices. This potential processability and scalability in the fabrication of large-area molecular electronic devices are much more appealing for industrial processes opening the door towards industrial mass production. Here, it is encouraging to note that a molecular electronic device for audio processing has already been commercialized [72], albeit briefly. Therefore, the translation of such science to a viable technology for industrial applications is an active goal today, with a considerable number of scientific and technological challenges [73] remaining to be met, as indicated below. Such incorporation of molecules into hard electronics will probably take place, at least in a first stage, in combination with traditional silicon-based current technologies. An optimization of interface engineering has been recently named as "the most challenging issue that hampers the development of reliable molecular junctions" in a recent review published in Nature Reviews Physics [13]. Such technology should be able to optimize the molecule-electrode contacts oriented to mass production of high yield, robust, stable, scalable, and reproducible devices to be produced at a reasonable cost. The optimization of the electrode | molecule interface (Figure 1c) involves several aspects:

(i)   Coupling of the molecule to the electrode surface through the contacting group, which plays a crucial role and has prompted an extensive search for chemical groups that can effectively serve as molecular 'anchor groups' [25,28,35,74–78];

(ii)   Mechanical stability of the electrode molecular junction avoiding fluxional bonds [79,80]. Several strategies have been employed for this purpose, including the use of multidentate anchor groups [57,81], and multipodal platforms [82];

(iii)   Compromise between the mechanical stability and electronic coupling since a too strong interface coupling may result in the loss of electrical functionalities and also in poor gating effects in three-terminal structures. This compromise could be reached by the insertion of a spacer between the anchoring group and the conjugated skeleton within the molecular structure [13];

(iv)   Control of the geometry of molecules to avoid fluctuations, due to different orientations (and then a different distance for the electrons tunneling between the electrodes), which can be achieved by an exhaustive control of the surface coverage and molecular packing density [83].

(v)   Control of lateral interactions and aggregation effects in molecular assemblies. These lateral intermolecular interactions may have a decisive role in the electron transport properties of large-area devices based on π-conjugated materials [22];

(vi)   Deposition of the top contact electrode avoiding the formation of short-circuits and/or damage of the functional organic molecules in the monolayer [84,85].

Taking into account this scenario, consideration is now being given to device fabrication strategies (deposition of a monolayer onto the bottom electrode and deposition of the top contact electrode onto the monolayer) for the construction of large-area devices that could progress the integration of the concepts of single-molecule electronics towards viable large-area devices. The objective of this focused review, with no claims of completeness, is to describe the most widely used strategies for the deposition of a monolayer onto the bottom electrode, as well as an overview of the top-contact deposition methods explored so far.

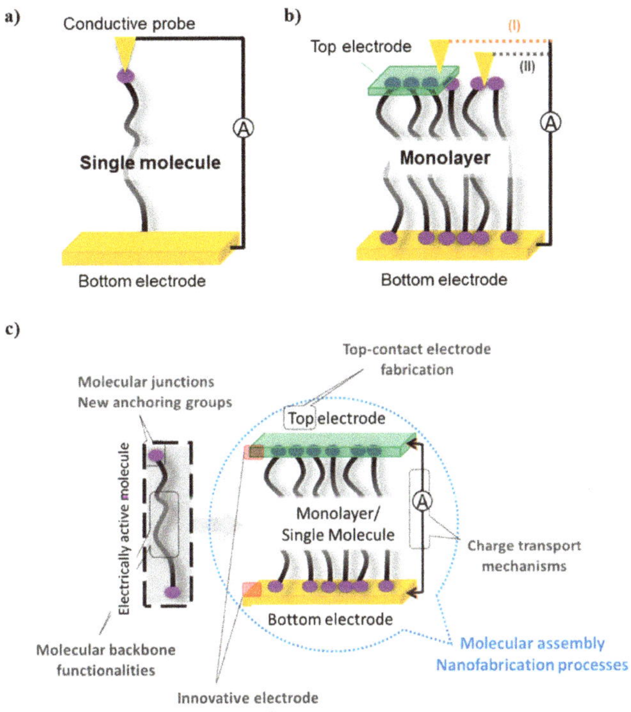

Figure 1. (**a**) Scheme of a molecular device based on single-molecule junctions. The electrical contact is made directly with a conductive probe. (**b**) Scheme of a sandwiched large-area molecular device. The electrical contact can be made using a top-contact electrode (I), which is more oriented to mass production, or directly above the organic monolayer (II), to determine the electrical properties of the electrode | monolayer structure. (**c**) Main challenges to be addressed in molecular electronics.

2. Fabrication of Molecular Films, Deposition Techniques

Nanofabrication of monolayers for the construction of vertical devices has attracted considerable attention in the field of molecular electronics, in particular by those working on the large-area device paradigm. An overview of the fabrication methodologies for these large-area methodologies is the scope of this section. A well-ordered monolayer sandwiched between two electrodes is the fundamental element to be studied in this field, with many seminal works establishing key structure-property relationships through the construction and study of electrode | monolayer | electrode system (molecular junctions). For this reason, bottom-up techniques—in which forces acting at the nanoscale are used to assemble molecules into large assemblies—have been widely employed for the deposition of a monolayer onto the bottom electrode and will be revised in this section. The most widely used techniques for the construction of electrode | monolayers in the context of molecular electronics include the Self-assembly (SA), the Langmuir–Blodgett (LB), and the Electrografting (EG) methods. Because of the remarkable ensemble capabilities of each one of these techniques, monolayers or multilayers made of organic, organometallic, hybrid inorganic-organic building blocks, as well as biomaterials have been deposited onto conducting or semiconducting substrates. In addition, the growing knowledge on the electrode | monolayer interface, together with the mature synthetic expertise in this research field, has boosted the ad hoc synthesis of materials with improved anchoring characteristics onto the bottom-electrodes [35,82,86–91]. Table 1, gathers, with no claim of completeness, some of the families of compounds that have been more widely used for fundamental studies in molecular electronics.

Table 1. Illustrative examples of molecular skeletons and anchor groups of the main families of materials used to fabricate molecular films in the field of molecular electronics.

	Molecular Structure	Anchoring Groups, Assembly Tech. and Reference
SATURATED HYDROCARBON CHAINS		X = Y = -SH_ n = 8–12_SA [92] X = -SH, -I,-C≡N, -N≡C, -Si(OCH$_3$)$_3$, -PO(OH)$_2$; Y = -H_n = 1–13_SA [93–98] X = -NH$_2$; Y = alkyl chains_EG [99] **MULTIDENTATED** X = -OPhCH(CH$_2$SH)$_2$; Y = H_n = 18_SA [100] X = -CH(CH$_2$SH)$_2$, -CH$_2$CS$_2$H; Y = -H_n = 14_SA [93] X = -PO(OH)$_2$; Y = -H_n = 12_SA [101,102] X = -SH; Y = -COOH_n = 8–16_SA [103] X = -CH$_3$; Y = -COOH_n = 20_LB [104]
	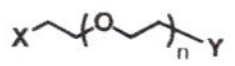	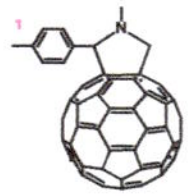 X = -PO(OH)$_2$; Y = -H_n = 2_SA [105] X = 1; Y= -H_ n = 4_SA [106]
	 X≡≡≡≡Y	X = Y = -H, Py, -PhCN, -PhNH$_2$, -PhSCOMe_SA [107,108]
	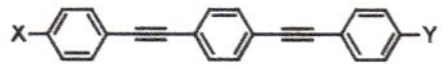	X = Y = -SH_SA [75,92] X = -H; Y = -SCOCH$_3$_SA [109] X = Y = -NH$_2$_SA [110,111] X = Y = -C≡CH_SA [112] X = -NH$_2$; Y = -O(CH$_2$)$_5$CH$_3$, -C$_5$H$_{10}$Si, -SCOCH$_3$_LB [104,113,114] **BIDENTATED** X = Y = -CS$_2$_SA [75] X = Y = -COOH_LB [115] X = -COOH; Y = -O(CH$_2$)$_5$CH$_3$, -C≡CH, -H _LB [116–120] X = -COOCH$_3$; Y = -OC$_{16}$H$_{33}$, -OC$_9$H$_{19}$_LB [121]
	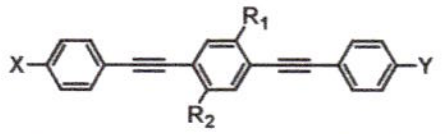	X = -H; Y = -SCOCH$_3$, R$_1$ = .NH$_2$, R$_2$ = .NO$_2$_SA [109]
		X = -CH$_2$SH; Y = -OCH$_3$, -NO$_2$, -H_SA [122,123]

Table 1. *Cont.*

	Molecular Structure	Anchoring Groups, Assembly Tech. and Reference
	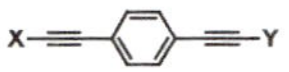	X = Y = -H_SA [124] X = Y = -Py_LB [125] **MULTIDENTATED** 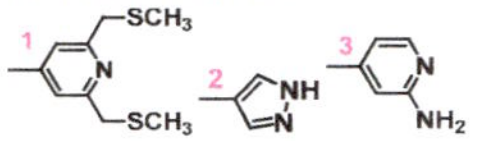 X = Y = 1_LB [126] X = Y = 2_LB [81] X = Y = 3_LB [127]
	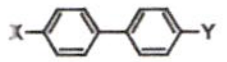	**MULTIDENTATED** 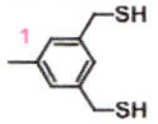 X = 1; Y = -OCH_3,-NO_2_ SA [122] X = -$PhCH_2N(CH_3)CS_2$; Y = -H_SA [123]
	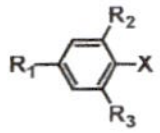	X = -N_2^+; R_1 = -Br, -NO_2, -CH_2OH, -$CH_2OSi^iPr_3$; R_2 = R_3 = -H_ EG [128,129] X = -I; R_1 = -H; R_2 = R_3 = -CH_3_EG [130] X = -NH_2; R_1 = -CH_2COOH; R_2 = R_3 = -H_ EG [129]
CONJUGATED MOLECULAR BACKBONES	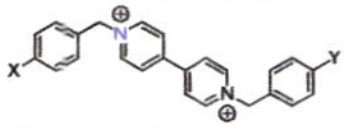	X = Y = -$C_5H_{10}Si$; -N-_LB [83]
	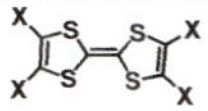	X = -SH_SA [131]
	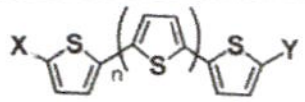	X = Y = -H; -S-_ n = 4_SA [132] X = -PhC_6H_{13}; Y = -$Ph(CH_2)_{12}SiCl_3$; -S- _ n = 0_SA [133]
	MULTIPODAL 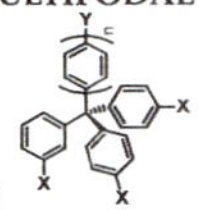 **(a)** 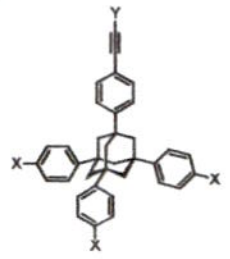 **(b)**	**(a)** X = -SH; Y = -CN_ n = 1–4_SA [134] **(b)** X = -$S(CH_2)_7CH_3$, -$S(CH_2)_{11}CH_3$, -SAc; Y = -$Fe(C_5H_5)_2$, -PhN_2PhCN_SA [135]

Table 1. *Cont.*

	Molecular Structure	Anchoring Groups, Assembly Tech. and Reference
ORGANOMETALLIC COMPOUNDS	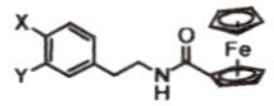	X = Y= -OH_SA [136]
	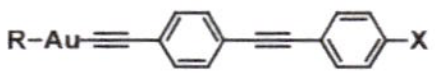	X = -NH$_2$; R = -CNC$_8$H$_9$_LB [21] X = -COOH; R = -PPh$_3$_LB [137]
	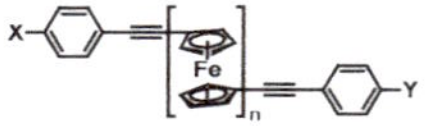	X = Y = -SH, -SCOCH$_3$_n = 2–3_SA [31]
	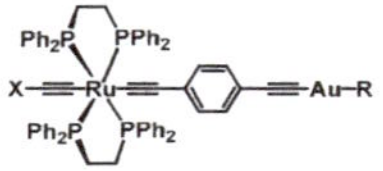	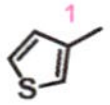 X = 1; R = -PPh$_3$ _SA [138]
	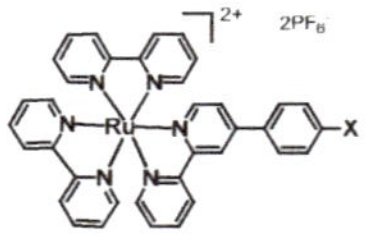	X = -NH$_2$_EG [139]
PORPHYRINS	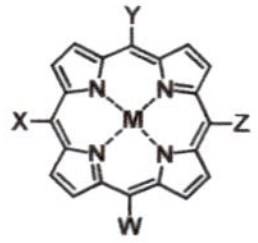 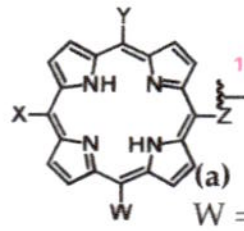	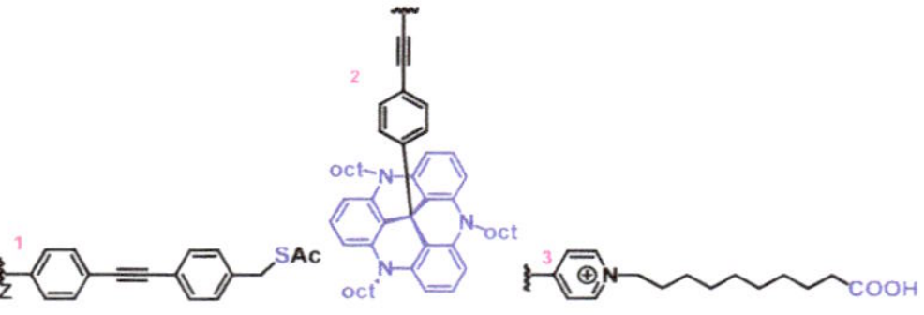 W = X = Y = Z = 1; M = Zn_SA [140] W = 2; X = Y = Z = -H, -Ph; M = Zn_SA [141] W = X = Y = Z = Py; M = TiO_LbL [142] **(b)** W = -NH$_2$; X = Y = Tol; Z = -Ph_ EG [143] W = X = Y = Z = 3_SA [144]

Table 1. *Cont.*

	Molecular Structure	Anchoring Groups, Assembly Tech. and Reference
CARBONACEOUS MATERIALS	**(a)** CNTs **(b)** rGO	**(a)** LS [145] **(b)** LB [146]
BIOMOLECULES	**(a)** DNA **(b)** Proteins	**(a)** LB [147] **(b)** SA [148,149]

2.1. The Self-Assembly Technique

The self-assembly technique is based on the autonomous and spontaneous reaction of a certain functional group on a surface (chemisorption) [150] and subsequent organization of the backbone of the molecule, due to non-covalent lateral intermolecular forces between neighbor molecules, resulting in long-range molecularly ordered domains within the monolayer. The cooperative effect of this strong-soft combination of forces often results in dynamic behavior of the molecules in the self-assembly monolayer (SAM), exhibiting lateral diffusion, conformational isomerism, and even reconstruction [151].

A general procedure for SAMs fabrication is illustrated in Figure 2. The substrate (e.g., Au, Ag, Pt, Cu, Pd, SiO_2, etc.) is introduced in an organic solution containing the molecule of interest (typically in the 10 µM to 10 mM range). A previous thorough cleaning process of the substrate is required, and often, a pre-treatment of the surface is also applied (annealing, plasma, etching, etc.). After a certain incubation time, the sample is withdrawn from the solution and exhaustively rinsed to remove any physisorbed material. In order to obtain high-quality monolayers with a large surface coverage, it is necessary to optimize a number of influential parameters, such as the incubation time, the concentration, the solvent, the temperature or the ambient relative humidity, due to the water adsorption on the surface of the substrate prior to the immersion into the solution containing the molecule of interest [152,153].

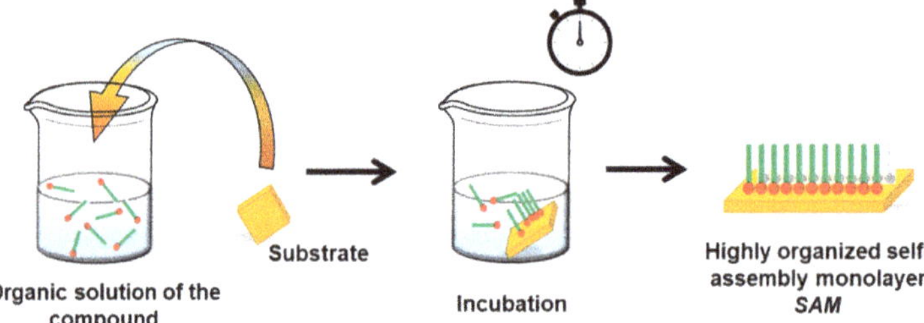

Figure 2. Scheme that illustrates the procedure for the preparation of a self-assembled monolayer. First, a clean substrate is incubated in a solution containing the molecules of interest for a certain time. Second, the substrate is withdrawn and thoroughly rinsed to remove the physisorbed material.

The self-assembly method has been widely used [154–156] in the field of molecular electronics since it is probably the most useful approach for the fabrication of well-organized and tightly-packed monolayers, due to its simplicity and versatility. The main advantages of the self-assembly methodology include:

- Fabrication of high-quality monolayers exhibiting 2D-crystalline long-range molecularly ordered regions.
- Low-cost process: (i) No specific instrumentation is required; (ii) it takes place under ambient conditions (no annealing neither low pressures required); (iii) no significant contamination problems occur upon the monolayer fabrication; (iv) the process is quite straightforward for the operator.
- The SA method is easily scalable.
- The assembly process can be in situ followed by a quartz crystal microbalance, QCM, that provides information about the deposition rate and surface coverage.
- Further functionalization of the monolayer either to form multi-layered films [157] or to deposit the top-contact electrode trough a strong molecule-electrode interaction is possible by an appropriate design of the material with the inclusion of a terminal group with the desired functionality.

Chemisorption is highly specific in nature, and therefore, the terminal functional group of the organic moiety needs to be carefully designed to find a compromise between the strength of the anchoring process on the surface and the final performance of the electrode | monolayer junctions [158].

In the self-assembly methodology, the main limitation comes from the number of functional groups showing the specific affinity and robust interactions with the electrode. As a consequence, a vast majority of the SAMs for molecular electronics incorporate molecules having a thiolate derivative as the head group interacting with the metal substrate [159–175]. In fact, most of the pioneering work in the study of electrical properties of molecules done in the late 1990s and early years of the 21st century was based on thiol on gold contacts [176–179]. These seminal works served to set the basis of the methods to measure the electrical properties of both single-molecules and ensembles of molecules and also to demonstrate the viability of using molecules as basic elements in circuitry. However, the Au–S bond is fluxional and not very stable, which explains some of the reproducibility problems encountered in the literature [92,109]. Additionally, other drawbacks to the thiol on gold SA technique have been reported, such as the tendency of organic thiols to oxidize to disulfides; this problem can be overcome by using protected thiols, but the incorporation of extraneous material within the system when in situ deprotection steps are involved may also affect the final electrical properties of the system [180]. The chemical reactivity and thermal stability of these systems in ambient and aqueous solutions have also been reported to seriously limit the technological applications of thiol and dithiol monolayers on gold [161]. Moreover, a recent contribution has revealed that SAMs prepared from the solution deposition of dithiols do not have a chemisorbed character [181]. For these reasons, many other anchor groups have been studied in the last two decades on different substrates (SiO_2, Cu, Ag, Au, WO_3, and ITO–indium tin oxide-among others), including selenols [94], amines [110,111], cyanides and isocyanides [95,96], isothiocyanates [182], trimethylsilyl group [183,184], acetylenes [107,112,124,185], thiophenes [132,186], trichlorosilanes [133], trimethoxysilanes [97,142], phosphonate [98,105] terminated molecules, perylenes [187], and fullerenes [106,188,189]. Recently, Qiu et al. [106] reported the spontaneous formation of molecular junctions of glycol ether functionalized fullerenes on Au^{TS}, which resulted in the large stability and robustness of the SAMs.

As it was mentioned before in the introduction section of this review, much interest has been paid recently to the use of SAMs incorporating multidentated anchor groups [190] that are expected to provide additional robustness to the molecular junction, lower fluctuation defects, more efficient electronic coupling, and enhanced electrical performance [26,170]. In this context dithiols [100,122], carboxylic acids [103], dithiocarbamates [123,191], carbodithiolates [75], dithiocarboxylic acids [93], tetrathialfulvalenes [131], phosphonic acids [101,102,192] and catechols [136] terminal groups have been assembled in SAMs and their electrical response evaluated. Some of these multidentated anchor groups have been demonstrated to exhibit superior electrical properties. For instance, dithiocarbamates on gold result in improved stable and low contact resistance junctions in comparison to thiol contacts, with a drop in the contact resistance by ca. 2 orders of magnitude [123]. Multipodal platforms also provide firm coupling between the molecule and the electrode through individual anchor points, and are also receiving increasing interest. Examples include tripodal [134,135,193] and tetrapodal [194] platforms incorporated onto SAMs which, allow to make a strong contact and to enforce an orientation of the molecules at a fixed distance from the surface [190]. Additionally, a selection of the bulky tripodal platform guaranties an effective separation with the metal surface, avoiding the quenching of the excited state caused by the metal surface in a photoisomerization process, allowing to develop optoelectronic devices of great interest in the electronic industry [82]. Construction of multilayer films using multipodal molecules by hydrogen-bond formation through pH control has also been used to demonstrate long-range electronic transport [192]. Other relevant properties in multipodal platforms have also been found, including attenuation of tunnel currents more effectively than do the corresponding monodentate SAMs, which may be useful in future applications for gate dielectric modification in organic thin-film devices [101]. More examples of both multidentate [74,81,195] and multipodal [196–198] platforms have been studied at the single molecular level, though the extension of these investigations to large-area ensembles is a topic of growing interest.

The incorporation of compounds with different molecular lengths has been recently explored as a tool to tune the electronic interaction between neighbor molecules in these mixed SAMs, which results in changes in the quantum tunneling performance of the devices [199,200]. This result opens the door to the exploration of new functionalities taking advantage of the lateral supramolecular organization of the molecules in SAMs. Whilst, the incorporation of photochromic moieties to fabricate switching molecular electronics devices, where it is possible to control the on/off state by external stimuli, is also a topic of interest [201].

In addition to the standard SA method, the combination of SA capabilities with other techniques, such as electrochemistry results in the electrically assisted self-assembly methodology [202–207] that may result in a significant decrease in the deposition time and an improve in the quality of the monolayer.

Self-assembly strategies in which molecules are located onto a gold electrode by forming covalent Au–C σ-bonds [208] have resulted in SAMs with a significantly higher conductance than those with the above-described conventional anchoring groups [76,209–211]; electrically transmissive monolayers with Au-C junctions have also been reported [112]. These results are attributable to the creation of exceptionally stable SAMs (~4 eV) and strong electronic coupling because of an uninterrupted conjugation between the electrode and the molecule [107,212]. These investigations have been extended to other interfaces, including various metal-C and C (from the electrode)-C (from the molecule) [212], as well as silicon-C junctions [34]. Ultrastable SAMs with high thermal, hydrolytic, chemical, oxidative, and electrochemical stabilities of N-heterocyclic carbenes (NHCs) on gold have also been demonstrated [213].

Proteins are basic elements that work as building blocks in bioelectronics devices, including protein-based transistors or sensing applications (monitoring bio-molecular interactions between target molecules and proteins). In this context, an important body of research in the field of molecular electronics is being done with proteins both at a single-molecule level [214–218] and thin-films comprising a monolayer or a short number of layers [219–223]. Additionally, it has been experimentally and theoretically proved that it is possible to tune molecular electron transfer rates in electron transfer proteins (ETpr's) through (i) chemical modifications and changes in the redox center, as well as the locations of the donor, the acceptor, and the bridge moieties within the ETpr's structure [220,224–228] (ii) modifying the solvent environment [229], and importantly, (iii) orientation relative to the electrode and changing the strength of the protein–electrode coupling [230–232]. The study of large-area protein-based molecular electronic devices has been carried out mainly by the self-assembly method [148,149] via an appropriate linker. DNA has also focused the interest of researchers in the molecular electronics community due to several reasons, including (i) superior self-assembly properties [233,234]; (ii) its unique electrical properties [235], and (iii) its potential use in bioelectronics devices. Additionally, a combination of self-assembly and dielectrophoresis (DEP) methodologies has been pointed out as an efficient tool in the construction of nanodevices [236]. Relative large materials in the field of molecular electronics, including proteins, DNA, carbon nanotubes, graphene oxide, nanoparticles, etc., can be assembled onto an electrode by means of DEP [148,223,237–243]. DEP is based on the movement of a polarizable nanoscale object (neutral or charged) caused by the polarization of such nano-object induced by a non-uniform electric field [244–252]. Coulomb interactions between the induced surface charges of the nano-object and the electric field occur. If the nano-object is in a uniform electric field, the net force acting on the nano-object is zero. In contrast, in a non-uniform electric field, there is a net force acting on the nano-object. This net force results in the motion of the nano-object since one end of the dipole is in a weaker field than the other, and the nano-object is pulled electrostatically along the electric field gradient. There are two types of DEP: positive and negative. If the nano-object experiences a force towards the high-field intensity region, the phenomenon is known as positive dielectrophoresis. In positive DEP, the nano-object has a larger polarizability than the surrounding medium, and it is pushed towards the region of a higher electric field. In contrast, if the nano-object experiences a force towards the low-field intensity region, the phenomenon is called negative dielectrophoresis. In negative DEP, the nano-object has smaller polarizability than the surrounding medium. Deposition through dielectrophoresis can be experimentally controlled by adjusting a series of parameters, including the dielectric constant of the nano-object and its surrounding medium, magnitude and frequency of applied electric field, and electric field gradient.

As mentioned in the introduction, this vast body of research related to SAMs of organic, organometallic, and biomaterials may find direct use in the incorporation of molecules in large-area devices that may work as molecular wires, diodes, molecular switchers, rectifiers, single-molecule or protein-based transistors, etc. The control of the thermoelectric properties of SAMs (by using an appropriate anchor group, where the Seebeck coefficient can be boosted by more than an order of magnitude), represents a critical step towards functional ultra-thin-film devices for future molecular-scale electronics [253,254]. Other applications of SAMs beyond the field of molecular electronics include, just to mention a few examples, the protection of electrodes from otherwise highly detrimental environments preventing electrochemical corrosion. This is a useful finding to increase, for instance, the lifetime of electrochromic devices [102]. Moreover, nanotemplating (which resembles the host–guest interaction in supramolecular chemistry) can be applied in molecular separation, chemical sensors and nanoreactors [255]. This technique has been also used as a tool to fabricate hybrid organic/inorganic nanostructures with application in sub-5 nm bottom-up patterning nanolithographic processes, which is an important point for the development of future electronic devices [256]. Other research fields where SA appears as a useful and versatile working technique include the development of drug delivery systems [257,258], the merge of advanced nanomaterials and optical fibers, known as lab-on-fiber optrodes [259], as well as biopharmaceutical applications [260], among many others.

2.2. The Langmuir–Blodgett Technique

In contrast to the SA method, in which molecules are first assembled onto the substrate, and subsequently, they are organized, in the Langmuir–Blodgett method, molecules (amphiphilic in nature) are first organized at the air-liquid interface (the liquid is usually water). The driving forces operating for the arrangement of the molecules at the air-liquid interface are mainly van der Waals interactions, hydrogen bonding, and/or electrostatic interactions. In addition, the pH of the liquid subphase, as well as the presence of ions can also govern the organization of the molecules [261]. Once the monolayer at the air-water interface, Langmuir film, is well-formed, it can be transferred onto a solid substrate to fabricate a Langmuir–Blodgett (LB) film. The transference can be done either by the emersion or the immersion of a vertical substrate (electrode) with respect to the air-water interface or by the horizontal lifting of a substrate located parallel to the subphase (Langmuir-Schaefer, LS, methodology). Depending on the nature of the solid substrate and the anchor groups present in the molecules, the molecules can be physisorbed or chemisorbed [126]. The main advantages associated with the Langmuir–Blodgett technique include:

- Fabrication of high-quality monolayers with high internal order;
- Fabrication of homo and heterogeneous multi-layered systems [262], resulting in highly ordered 3D molecular architectures;
- Fabrication of directionally oriented monolayers when asymmetric molecules are used [37,114].
- Large control of the orientation and the packing density of the molecules within the LB monolayers through optimization of different parameters that can be modified upon the manufacturing process, including the nature of the subphase, the spreading solvent (or even mixture of solvents), the temperature, the closing barriers speed, the dipping speed, the transference pressure, or transference direction of the electrode;

- *In situ* characterization of the Langmuir film upon the compression process by a wide variety of techniques [263–270] as surface pressure vs. area per molecule and surface potential vs. area per molecule isotherms, Brewster Angle Microscopy (BAM), ellipsometry, X-ray reflectometry, dilational rheology, Infrared reflection spectroscopy or UV-vis reflection spectroscopy, etc. These techniques provide complementary information for the understanding of the intermolecular interactions in the film;

- A really large number of different anchor groups can be used in the LB methodology in contrast with the SA method since these groups can be not only chemisorbed but also physisorbed onto the electrode. Furthermore, the transference of these films is possible to almost all types of substrates (e.g., metals, conducting polymers, silicon, carbon-and graphene-based electrodes, etc.).

The historical background and comprehensive descriptions of the technique have been reported before [90,91,261,271,272]. The Langmuir–Blodgett technique, illustrated in Figure 3, involves a more complex monolayer formation process compared to the SA method. Briefly, a Langmuir though is used. The basic elements of a Langmuir though include a cuvette made of a hydrophobic material, one or two barriers, and a Wilhelmy plate to determine the surface pressure (defined as the difference between the surface tension of pure water minus the surface tension of the water with the monolayer). Additionally, the transference of the monolayer onto the substrate requires a dipper to introduce or withdraw the substrate. The starting point in the LB process is the spreading of a solution of the molecule in an organic solvent. The solvent must dissolve the material, be highly volatile, and exhibit a large spreading coefficient. A certain volume of this solution is carefully spread onto the subphase. After waiting for the solvent evaporation, the compression process starts with the aid of one or two mechanical barriers, and the surface pressure vs. area per molecule is recorded. Therefore, as the available area per molecule is being reduced, the molecules get closer resulting in intermolecular interactions. This gradual change in a molecular arrangement is reflected in surface pressure variations registered with the Wilhelmy balance. Thus, upon the compression process, the monolayer undergoes several bi-dimensional phases (gas, liquid expanded, liquid condensed, and solid) and phase transitions. Once the monolayer reaches the target surface pressure of transference, the Langmuir film can be deposited onto a solid substrate, forming an LB monolayer (LBM).

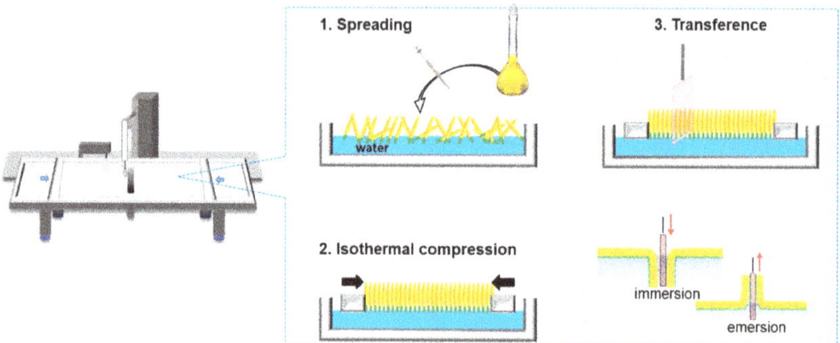

Figure 3. Scheme showing the main steps involved in the Langmuir–Blodgett monolayer (LBM) fabrication. First, an organic solution containing the molecule of interest is spread on the water surface. Second, the molecules are organized upon an isothermal compression resulting in a Langmuir film. Finally, the monolayer is transferred onto a solid support—forming the LBM.

Possibly the main drawback of the LB method is that it is a time-consuming technique. Concerning the fabrication process, a slow and critical step is the manual spreading process. Here, the electrospray (ES) spreading methodology could be an alternative to make this technique more functional and likewise to improve reproducibility. For instance, this methodology has been recently employed by Hirahara et al. [273] to obtain LBMs incorporating clay mineral nanosheets hybridized with the ODAH$^+$, where a comprehensive analysis of the influence of the infusion rates on the film properties is carried out.

In general terms, the materials typically used for the fabrication of LBMs are water-insoluble molecules with a head (hydrophilic)-tail (hydrophobic) amphiphilic structure [90,262,271,274]. Molecular assemblies using the LB technique for fabricating large-area molecular electronic devices of highly conjugated materials with monodentated anchor groups, such as thiols [114,275], nitriles [79], amines [21,276], trimethylsilylethynyl group [113,277], acetylenes [120], pyridines [125,278], or a viologen moiety [83], have been constructed. Additionally, the use of multidentate or multipodal anchor structures to improve the robustness and stability of the molecular junction has also been explored. For instance, using acids [116,118,279] or diacids [37], methyl esters [121], tetrathiofulvalenes [280], a 2-aminopyridine group [127], a pyrazole moiety [81], or the tripodal head group 2,6-bis((methylthio)methyl)pyridine [126].

Conductance in large-area molecular electronic devices incorporating carboxylic groups with lateral intermolecular H-bonding can be enhanced by deprotonation in a basic media with the subsequent rupture of those H-bonds. Therefore, this demonstration exploits the switching behavior in the conductance under protonation–deprotonation conditions for the construction of pH sensors based on molecular junctions [37]. Moreover, an LB film of a dyad consisting of an electron-rich "π-donor" (D) and an electron-poor "π-acceptor" (A) separated by a rigid, insulating "spacer" (featuring three main components of the original Aviram–Ratner rectifier design) has shown rectification behavior validating the rectification proposal by Aviram, and thus, allowing the possibility of producing organic molecular rectifiers [280].

In addition to the typical amphiphilic molecules described above, the LB technique is largely enlarged towards the assembly of more complex structures with interest in molecular electronics (and organic electronics). Polymers [281–283], metalloporphyrines [284], DNA [285], perylenes [286], perylene–NH$_2$ [287], ruthenium complexes [288], organo-modified inorganic nanoparticles in combination with polymer nanospheres (nano-mille-feuille system) [289], pillar[5]arene derivatives [290], polymer-coated CsPbBr$_3$ nanowires [291], aligned SWCNTs [145], semi-conductive 2D materials based on 2,3,6,7,10,11-hexaaminotriphenylene (HATP) [292], or rGO [146] have been used.

Although this review is focused on the potential of this bottom-up technique in the construction of monolayers with applications in molecular electronics, the LB technique has been applied in many other research fields. Here, it is worth mentioning the role of the LB methodology in many other fields of research that cover a large in the emerging concept *nanoarchitectonics* [293,294], which combines nanotechnology with other fields, such as supramolecular chemistry, nano/micro fabrication, organic chemistry, and bio-related technology. Nanoarchitectonics represents today a promising and powerful strategy in which the LB method yields a perfectly molecular organization within a 2D plane [295,296]. Therefore, because of the wide generality of the nanoarchitectonics concept, LB films can also be applied to a wide range of research fields with practical importance, such as materials production [297–299], sensing [300,301], catalysis [302,303], device [304,305], energy [306–310], or biological/biomedical applications [311–314], or even in the fabrication of smart textile-based sensors (TEX sensors) [311]. These studies underscore the almost limitless possibilities of the LB technique to fabricate well-ordered 2D films of a wide range of materials.

2.3. The Electrografting Technique

Electrografting (EG) is a well-known approach for surface functionalization or modification, in which an electrochemical reaction takes place between the conductive substrate and organic material [312]. This methodology has two main and powerful advantages:

- ■ The formation of a direct covalent bond between the electrically active molecule (s) and the electrode, resulting in stable and robust molecular junctions.

- Carbonaceous electrodes (in addition to other conductive substrates, such as metals, metal oxides, polymers, and semiconductors) are relatively easy to functionalize. The use of carbon-based electrodes represents a growing trend today in the development of sustainable technology able to manufacture electronic devices free of expensive and contaminant materials [313].

Nevertheless, the main drawback associated with this methodology is the tendency to form non-ordered multi-layered systems, due to the extremely high reactivity of free-radicals involved in the electrografting process. Several strategies have been successfully used in order to avoid such a multilayer growing. These approaches include molecular functionalization with bulky groups or the use of redox mediators or inhibitors [314,315].

EG covers reductive and oxidative processes, depending on the organic compound, as was comprehensively described before by Bélanger and Pinson [312]. Since this methodology was developed [316], different functional groups have been used for surface modification, including aliphatic amines [99], aromatic amines [317], or diazonium salts [314,315].

In the field of molecular electronics, widespread research concerns the modification of a bottom electrode (metals, carbon-based electrodes, hydrogenated silicon surfaces, etc.) by electrografting of diazonium salt derivatives [314,315,318,319]. One of the preferred methodologies to generate diazonium cations is the in situ diazotization of aromatic amines, as illustrated in Figure 4 [320]. As a general procedure, the diazotization step is followed by the electrochemical reduction, where highly active aryl radicals are produced. Subsequently, these radical intermediates covalently bind to the surface [128].

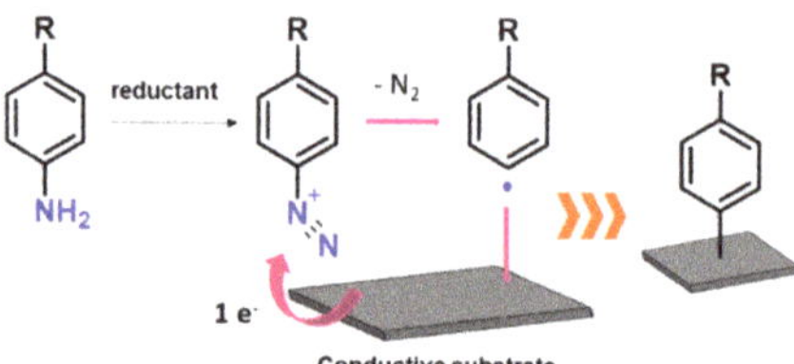

Figure 4. Scheme showing the electrografting process for a diazonium salt.

Numerous examples can be found in the literature reporting molecular electronic devices based on the electrografting of diazonium salts. Here, it is noteworthy the pioneering work done in the group of McCreery [321,322], that reported the first all-carbon based molecular tunnel junctions. This group also developed methodologies for the grafting of diazonium salts onto a pyrolyzed photoresist film (PPF) [323] and has recently reported an all-carbon molecular electronic device constructed on flexible or semi-transparent substrates [33]. These methodologies and materials pave the way for applications in which molecular electronics and photonics are combined. Additionally, the McCreery group also launched the first commercialized molecular electronic device, in which a molecular rectifier composed of a molecular layer sandwiched between two carbon-based electrodes is used for audio processing applications [72].

A different and emerging family of suitable molecules for indirect or direct electrografting are iodonium salts [130]. Concerning the direct grafting, Guselnikova et al. [324] have reported the surface functionalization of a gold substrate by UV-light grafting of the 3,5-bis (trifluoromethyl)phenyl)iodonium salt. Ramírez-Chan et al. [325] recently described the fabrication of electrografted films by oxidation of a nitrophenylbutyrate derivative ($NO_2Ph(CH_2)_3COO^-$) exploring the influence of different supporting electrolyte ions on the film formation. Madsen et al. [326] used a two-phase bipolar grafting system to simultaneously functionalize gold bipolar electrodes with diazonium salts and primary amines or thiophenes.

In addition to the fabrication of exceptionally robust monolayers for molecular electronics applications, this technique is an affordable methodology to reach specific improvements in surface functionalization for an extensive variety of substrates. For example, mixed monolayers made of organophosphonic derivatives have been electrografted onto nitinol (NiTi), which is an alloy often employed in the biomedical field, in order to prevent the release of the possibly carcinogenic Ni^{2+} ions [327]. Kim et al. [143] modified a GLAD-ITO substrate by the electrografting of a porphyrin diazonium salt, showing the possibility to introduce molecular functionalities, such as photo-activity on nanostructured metal oxides electrodes. Moreover, polymers, such as the PEDOT (Poly(ethylenedioxythiophene) films, have been electrografted to repair IMPC (Ionic exchange polymer metal composite) to overcome the problem of water leakage and also providing reinforced properties to the electrode [328]. Another approach to enhance the adhesion of PEDOT onto ITO and gold surfaces was presented by Villemin et al. [329], who introduced a two-step strategy in which the first step consists of the fabrication of an electrografted promoter layer based on EDOT or thiophene moieties. Likewise, materials, such as carbon nanodots (CD) have been immobilized onto carbon substrates, through the attachment of nitrogen moieties at the nanodot surface (diazotized N-CD), resulting in carbon substrates with new performances in sensing applications [330]. Finally, other conductive substrates such as carbon composite electrodes, including thermoplastic electrodes (TPE) [331], edge plane pyrolytic graphite electrodes (PGEs) [332], glassy carbon (GC), and pyrolyzed photoresist (PPF) [333], have also been modified by electrografting.

Finally, two useful and recent advances in this field are worth mentioning. One of them results from the merge of the LB and the electrografting techniques and was recently published by Gabaji et al. [334]. In this methodology, the electrografting process occurs simultaneously to the transference of the Langmuir film onto the solid substrate. The other relevant approach results from the combination of EG with the electrode mediated shadow edge deposition methodology; here, molecules of 9,10-dioxo-1-anthracenediazonium salt were covalently attached to metallic nanotrenches producing stable and reproducible lateral architectures of molecular junctions [335].

Although EG is widely utilized for the construction of monolayers with applications in molecular electronics, this technique is also used in many other research fields as well as in industrial applications. In this context, it is worth mentioning the development of a diazonium-based biochip for surface plasmon resonance (SPR) analysis, a well-established technique for studying affinity between biomolecules whose interaction takes place in a liquid/solid interface [336], by electrografting of a carboxybenzene diazonium salt [337]. Additionally, electrografting, followed by a post-functionalization, is being an interesting approach in the design of well-defined interfaces for electrochemical (bio) sensing, very important in fields, such as chemistry, materials science, engineering, biology or medicine [320]. Furthermore, EG is an important tool to modify electrodes in bipolar electrochemistry applications, which involves two feeder electrodes and a conducting object (the bipolar electrode) in an electrolytic solution; a field of renewed interest in the last decades, due to its use in materials science or sensing [338]. Accordingly, EG can be applied in a wide range of applications in the areas of catalysis, biosensors, sensors, corrosion protection, composite materials, energy conversion, energy store, or superhydrophobic coatings, which reveals the enormous possibilities of the EG methodology to fabricate bi-dimensional arrangements using a wide range of materials. EG is also used in industrial applications, such as the modification of carbon black by simple mixing with diazonium salts generated in situ in aqueous solution (patented by Cabot Coorporation, Boston, MA, USA) to produce pigments for inkjet applications, automotive coatings, solar cells, and fuel cells [339]; or the fabrication of drug-eluting stents (endovascular devices to deliver locally therapeutic agents) [340]; revealing that EG technique is as a powerful tool for real-life applications.

To sum up, in this section, the most relevant techniques for the construction of monolayers onto (semi-)conducting or dielectric surfaces for molecular electronic applications have been covered, with examples of a large variety of technological applications. Table 2 gathers the main concepts here presented and may serve as a comparison scheme with the different pros and cons of each of these methodologies.

Table 2. Comparison of the deposition techniques to fabricate molecular films.

	Method		
	SA	LB	EG
Type of Adsorption	Chemisorption	Chemisorption Physisorption	Formation of strong chemical bonds between the molecule and the electrode
Process	Adsorption comes first, molecular organization follows	Molecular organization comes first (air-liquid interface), adsorption follows	Deposition comes first; the subsequent molecular organization is rather limited
Surface bond energy	~1.9 eV (most of monodentated anchor groups on metals, metal oxides, etc.)	~1.9 eV (chemisorption)	~3.5–4 eV
	~4 eV (combination of anchor group + substrate leading to metal-C, C-C, Si-C, Si-O-C, Si-Si, Si-O-P, etc.)	~0.5 eV (physisorption)	
Types of substrates	Metals, metal oxides ITO, HOPG, graphene, Si, SiO_2, etc.	Any substrate (if physisorption occurs; for chemisorption metals, metal oxides, ITO, C, and Si-based electrodes)	Conductor or semiconductor substrates (metal, C-based electrodes)
Advantages	■ High-quality monolayers ■ Low-cost process ■ Scalable ■ In situ characterization of surface coverage ■ Further functionalization of the monolayer is possible (multilayers, top-contact)	■ High-quality monolayers and multilayers ■ Directionally oriented monolayers ■ Large control of the orientation and the packing density ■ Scalable ■ In situ characterization upon the fabrication process ■ Large number of molecule-substrate interfaces	■ Formation of a covalent bond between the molecule and the electrode: stable and robust molecular junctions ■ Carbonaceous electrodes can be functionalized
Limitations	Rather limited number of (anchor groups + substrates)	Time-consuming technique	■ Multilayer formation ■ Very strong molecule-electrode bonds that may be a limiting factor in the effectiveness of any subsequent molecular organization on the surface

3. Fabrication of the Top Contact Electrode

Once the molecules have been assembled using some of the methodologies above-described onto a bottom-electrode, the subsequent deposition of a top-contact electrode to close the circuit is one of the most crucial and delicate steps in molecular electronics. Even when the deposition of the top-contact electrode has been largely investigated, having a reliable control in the fabrication of this electrode is still a challenge to reach viable incorporation of molecular electronics in the market. Different techniques have been employed to deposit or make a top-contact electrode, including physical or chemical vapor deposition, atomic layer deposition, liquid metal droplets, break junctions, scanning probe tips, electrodeposition, electroless deposition, etc. Moreover, these techniques can be divided into two categories: (i) Those that can be used to integrate molecular constructs into devices; and (ii) those that are employed for characterizing molecular electronic properties. In this review, we will focus mainly on the first ones. Moreover, the reader can find elsewhere excellent and comprehensive reviews that describe in detail the methods used to characterize, in the laboratory, molecular electronic properties through the formation of temporal metal-molecule contacts, but that are unsuitable for large-scale integration such as: In situ break junction (STM-BJ), mechanically controlled break junction (MCBJ), $I(s)$ (I = current, s = distance) and $I(t)$ (I = current and t = time) methods based on an STM, electromigration breakdown junction, liquid metal droplets, etc. [9,84,169,341]. In the MCBJ, firstly, a fine metal bridge is formed, and subsequently, this is cleaved upon bending the whole assembly allowing forming the molecular bridge. Similarly, in the STM-BJ method, break junctions are mechanically formed using an STM tip to create a metallic contact to the substrate, which is then cleaved, forming the molecular bridge, while monitoring the current. On the contrary, in the $I(s)$ method, although it also uses an STM tip to form molecular junctions, there is no contact between the metal electrodes since the STM tip is brought close to the surface and then withdrawn, while the tunneling current is measured. Meanwhile, in the $I(t)$ method, the STM tip is placed at a constant distance from the substrate, and the formation of molecular wires is monitored in the time domain.

Top-contact electrodes on molecular assemblies have been fabricated by direct evaporation of metals using either physical vapor deposition (PVD) [342] or chemical vapor deposition (CVD) [343]. In PVD, the top-electrode is formed by evaporating a metal, at a sufficiently high temperature and low pressure, for the subsequent condensation of metal atoms in the molecular layer. Nevertheless, in this method, both the film thickness and the spatially selective growth of the metallic film are difficult to control. CVD is a more selective technique and permits higher control in the thickness of the metal deposition than PVD. However, CVD is characterized by rather slow growth rates unless high thermal activation temperatures ($\geq$200 °C) are employed, with these high temperatures being incompatible with most organic thin-films. Regardless if PVD or CVD are used, both methods often result in damage to the monolayer and also in penetration of metal atoms through the monolayer, and therefore, metallic contact between the top and bottom-electrodes [84,344,345]. Alternatives (such as metal evaporation onto a cooled substrate [346,347], blocking the direct path between the crucible and the sample with baffles [348], or the use of an indirect evaporation method to reduce the exposure of the molecular layer to energetic metallic atoms and temperatures [349]) have been proposed to minimize damage to the monolayer, as well as the presence of short-circuits.

The incubation of a functionalized monolayer into a metal nanoparticle dispersion results in the chemical deposition of the metal nanoparticles on top of the organic layer. In this context, gold nanoparticles and thiol functional groups have been widely used [350]. However, this methodology results in an incomplete metallization of the organic layer [350]. Therefore, alternatives to increase the surface coverage of the metal deposits have been developed. For instance, the photoreduction by UV-vis light of a metal precursor provides surface coverage of the monolayer surface as large as 76% [115,277]. In this methodology, the metal precursor is incorporated from an aqueous subphase to an LBM upon the transference process, generating in situ metal nanoislands on top of the monolayer by photoreduction of the precursor. This method has been proved not to damage the underlying organic film, nor penetration of the nanoparticles through the film Additionally, anisotropic palladium nanostructures, previously produced by a CO-confined growth method [351,352], were deposited onto the monolayer to generate large palladium nanodeposits (PdND) across the surface without any damage to the monolayer, and importantly, a surface coverage of 85% [111]. Moreover, the incorporation of in situ generated uncapped gold nanoparticles (NPs) has been explored. Here, the NPs were attached to a terminal alkyne functionalized monolayer through the formation of Au–C σ-bonds, via a heterolytic cleavage of the alkyne C–H bond [107].

An interesting methodology for the soft deposition of metallic contacts such as Au, Pt and Cu with yield >90% is the surface-diffusion-mediated deposition (SDMD) [353,354]. In this approach, firstly, a SiO_2 etch mask patterned on a pyrolyzed photoresist film (PPF) layer, is fabricated by optical lithography on a silicon substrate with a thermal SiO_2 insulating layer. After that, the molecular layer is formed on the conducting PPF substrate through a C–C bond by electrografting. In order to protect the monolayer from the source radiation or from the direct impingement of metallic vapor generated by using electron-beam evaporation, an adjacent silicon dioxide layer is used. Surface diffusion of the metal atoms from the silicon dioxide to the molecular layer leads to the formation of a "soft" top contact, albeit this approach has a significant addressability problem.

Atomic layer deposition (ALD) is a chemical process related to CVD but differs from it in several key aspects [355–357]. ALD involves a well-defined chemical reaction route to form the top-contact. In this methodology, the related chemical reactants (or precursors) are sequentially introduced into the reaction chamber via short pulses (each of them followed with a purge of inert gas to remove excess reactants). A layer-by-layer deposition onto the organic film takes place. Here, the free-ends of the molecule have been specially designed to exhibit affinity by these chemical reactants (or precursors) favoring a selective reaction. Since the film growth is self-limiting, atomic-scale control of the film thickness and minimization of the defects (pinhole-free films) is achieved. Subsequently, this protective molecular layer can be used to deposit a thicker metal film using traditional methods. ALD is very advantageous in fabricating metal-insulator-metal tunnel junctions (MIMTJ) and provides a high control of the thickness of the ultrathin insulating layer inserted in the junction [356]. For these reasons, ALD is considered an enabling technology for future electronics. In addition, based on this methodology, the conductor polymer PEDOT:PSS (where PSS is poly(4-styrenesulphonic acid)) has been used as an interlayer in large-area molecular junctions resulting in very stable devices with a shelf life of more than several months. The main drawback of this methodology may be associated with the presence of hydrophobic end groups in the monolayer that can difficult the subsequent deposition of the hydrophilic PEDOT:PSS [358].

Electrodeposition could be an attractive method for the metallization of monolayers because it does not require expensive vacuum equipment, and it is often easier to control. However, this methodology has so far been unsuccessful since the formation of clusters, or the penetration of metal ions from the solution or even metal wires through the defect sites in the monolayer are often observed—resulting in very low yield devices [84,359,360]. Although these problems—which are mainly associated with the existence of imperfections in the monolayer and the presence of free metal ions in the solution—have been able to be overcome or reduced [22,361–363], still exits serious limitations in this approach. Only 1/3 of the monolayer is covered by the top metal layer; albeit this value can be increased until an almost completely covered SAM by repeating the adsorption-electrochemical reduction cycle in a metal ion free solution, but with the inconvenience of increasing the presence of short-circuits [364,365].

An alternative to reduce or eliminate the inconvenient of metal penetration is electroless deposition. This approach allows the deposition of metals and other materials on a variety of substrates [84,366]. It is very similar to electrodeposition in the sense that the metal deposition results from the reduction of metal ions from solution, but without applying an external potential, which additionally makes this method compatible with insulating or low conductivity materials. However, electroless deposition often requires the use of a catalyst to be adsorbed on the surface prior to the metal deposition, which can contaminate the organic layer affecting the final functionality of the junction [367–369]. To control the diffusion of the catalyst, the organic layer can be functionalized to selectively adsorb the catalyst, limiting the metallization to the functionalized areas [370,371]. A modification of this method is the electroless nanowire deposition on micropatterned substrates, which is employed to direct the growth of metallic nanowires on the monolayer surface [368].

To solve the problems that arise from damage to the monolayer or formation of short-circuits, different non-destructive methods have been developed. Nanotransfer printing (nTP) is one of these methods that fabricate soft top-contact electrodes by transferring the metal contact to the monolayer (by mechanical contact) from elastomeric poly(dimethylsiloxane) (PDMS) or perfluoropolyether (PFPE) stamps [372,373]. The transference is a consequence of the chemical affinity between the metal contact and the anchoring groups present in the monolayer, for instance, a thiol-terminated monolayer and a gold metal contact [374]. By using the lift-off float-on (LOFO) method [374], a metal film can be transferred onto a monolayer via capillary interactions avoiding the physical damage in the monolayer. This approach includes basically four steps: (i) Evaporation of a metal film onto a solid support (denoted as leaf); (ii) detachment of the metal leaf from the solid support by floating the leaf at the liquid surface (lift-off); (iii) adsorption of a monolayer onto a solid support; and (iv) attachment of the metal film at the liquid surface to the monolayer supported in the solid substrate using a liquid-mediated process (float-on). Several different types of molecular junctions without any observable damage to the monolayer have been formed using this method [9]; however, the wrinkling of the leaf, the presence of air gaps between the monolayer and the leaf, as well as small monolayer-metal contact areas limit LOFO methodology for further applications and mass production. The transfer of a top metal layer with a hydrophobic polymer, polymer-assisted lift-off (PALO, which combines aspects of nTP with LOFO), solves the wrinkling problem and the small metal contact area in LOFO. Using PALO, metal electrodes with dimensions from 100 μm^2 to 9 mm^2 can be produced with high yield ($\geq$90%) [375]. In this methodology, the alignment of the top and bottom contacts without using cross-bar geometries is a significant difficulty.

Other relevant methods for fabricating molecular junctions are via the construction of a hole, such as a nanopore or a nanowell [375–377]. To fabricate a nanopore device, a thin Si_3N_4 layer is deposited in a double-sided polished silicon wafer. A window, suspended over the silicon wafer, is obtained by optical lithography, reactive ion etching (RIE), and wet etching techniques. After that, a gold layer is evaporated onto the top side of the membrane, filling the pore (the top electrode). Then, the sample is immediately incubated in a solution to assemble a molecular layer on the top electrode surface. Finally, the sample is placed in a vacuum chamber to thermally deposit the bottom metal (Au) electrode. Recently, large-area junctions based on SAMs deposited in AlO_x micropores fabricated on ultraflat template stripped bottom electrodes of gold (Au^{TS}) have been fabricated with high mechanical stability [378]. The construction of a nanowell is more simplified, and occurs in a planar device in contrast with the nanopore. In this case, a nanowell device is fabricated on silicon wafers incorporating pre-patterned gold electrodes covered with silicon dioxide prepared with the focus ion beam (FIB) technique. After the hole is created, molecules are self-assembled on this bottom electrode, and the hole is filled with gold to close the circuit. The disadvantages of these approaches are: The requirement of delicate processes during the sample fabrication, the non-reproducible electrical properties from sample to sample, and a higher ratio of short-circuits by penetration of the metal, overall in nanopore devices. However, to overcome these disadvantages, a conducting buffer interlayer can be inserted between the metal layer and the monolayer to avoid damaging the organic layer, which allows the creation of large-area molecular junctions [358,379–386]. Another alternative to fabricate large-area junctions is to create microfluidic channels made with an elastomeric polymer (PDMS) on the bottom electrode using conventional photolithography in which the organic layer is self-assembled, and then the channels are filled with Ga_2O_3/EGaIn electrodes [387]. It is noteworthy that even when large contact areas can be fabricated by LOFO, PALO, nano (micro) pore, or nanowell and microfluidic channels approaches, these techniques are not easily scaled up for the fabrication of molecular electronic devices; in contrast, nanotransfer printing approaches represent a promising methodology to fabricate integrated molecular devices.

Other radical different alternatives to fabricate "soft" top-contact electrodes have been developed recently. Examples include the so-called Thermal Induced Decomposition of an Organometallic Compound (TIDOC) method in which an organometallic film is thermally decomposed—resulting in the formation in situ of gold nanoislands on top of the film without producing short-circuits by penetration of the metal top-contact electrode through the layer [137,138]. Moreover, the fabrication of nanotrenches of controlled width by means of the shadow edge evaporation method resulting in molecular junctions with very good stability and reproducibility, which offer the advantages of large scale integration, reduced leakage currents, and easier access to the molecular layer with external tools [335]. Additionally, the DEP method (described before in this review) has also been used for the controlled deposition of a top contact electrode. Fereiro et al. used Au nanowires electrostatically trapped between two microelectrodes. The Au nanowire act as the top-contact electrode in a flat gold | protein monolayer | Au nanowire junction [148].

While the ordinary metal electrode molecular junctions described above remain operational, experimental platforms based on non-metal materials are being constructed, leading to new possibilities for molecular-scale electronics. Carbon electrode–molecule junctions are one of these new testing systems, where several carbon materials, such as reduced graphene oxide (rGO), graphene, or single-walled carbon nanotubes (SWNTs) are used as top-contact electrodes because of their unique advantages [9]. Graphene or rGO films fabricated by chemical vapor deposition have been widely used as the conductive interlayers in molecular junctions [379,384]. Meanwhile, soft top-contacts for the non-destructive fabrication of molecular junctions using floating-processed ultrathin graphene films [384], or transferred onto the layer via poly(methyl methacrylate) (PMMA)-mediated transfer [388], have also been implemented.

Nanoscale gaps in SWNTs with precise control are fabricated by lithography-defined oxidative cutting [389], resulting in SWNT-molecule single-molecule junctions, which could be easily extended to industrial mass production since all of the operations in the process are compatible with conventional micro- and nanofabrication techniques [9]. Meanwhile, using dash-line lithography [390], robust single-molecule junctions with a high yield, based on indented graphene point electrodes can be created. This methodology provides a large control regarding the size of the nanogaps through the regulation of the etching process.

Electron beam evaporation of carbon (e-C) to create soft top-contact electrodes onto monolayers has been used successfully with a high yield, excellent reproducibility, and thermal stability [33,321,391–393]. The fabrication of amorphous carbon top-contact electrodes (with well-defined shape, thickness, and precise positioning on the film) from a naphthalene precursor using a focused electron beam induced deposition (FEBID) technique has been recently employed to create molecular junctions with a high yield and stability [117].

Once the top-contact electrode is fabricated to close the circuit, the typical way to verify that these molecular junctions, created with a given structure, are reliable is to statistically investigate the electrical characteristics of these devices. Conducting-atomic force microscope (C-AFM) and scanning tunneling microscope (STM) are the techniques usually used to address the electron transport properties of these devices by registering current-voltage (I-V) curves (or density current-voltage (J-V) curves). A detailed analysis of these curves together with temperature-varying measurements allow to rule out the presence of short circuits; also to determine conductance through the fabricated molecular junctions, as well as establish the mechanisms that govern the charge transport; in addition, it is possible to elucidate energy level alignments through a spectroscopic analysis (by the transition voltage spectroscopy, TVS, method). Theoretical models are also employed to better understand the charge transport mechanisms through the molecular junctions. Whilst, the shape of the obtained conductance histograms for the molecular junctions is an efficient technique to determine the interaction energy between adjacent molecules, one of the key parameters to understand and optimize the performances of these large-area molecular electronic devices. Additionally, inelastic electron tunneling spectroscopy (IETS) can be utilized to investigate the vibrational modes in the molecular junctions. In addition, the thermoelectricity of the molecular junction, which measures the induced voltage drop or the induced current across the junctions between two electrodes at different temperatures, helps our understanding of the mechanism of the thermoelectric effect, and therefore, may serve to improve the technologies for converting wasted heat into useful electrical energy. Finally, the determination of durability and operational stability of the device is crucial for the practical application of these devices in the actual industry.

4. Summary and Outlook

The use of molecules in electronic devices as critical functional elements in circuitry is expected to result in a novel technology opening the path to future industrial processes for high-value products. The use of functional organic materials has attracted a great deal of attention due to the numerous and very appealing advantages of using molecules as functional units. Importantly, the use of molecules in electronic devices is expected to result not only in the further miniaturization of transistors, but also in diminished power consumption and new functionalities, due to the quantum effects that govern the properties at the nanoscale. Researchers are currently immersed in the exploration of two different paradigms. On the one hand, the study of single-molecule junctions, which is of primary importance to understand charge transport through molecules. On the other hand, the fabrication and characterization of large-area devices in which assemblies of molecules are located between two (or three) electrodes. The fabrication of these electrode | monolayer | electrode systems can be done through simple, soft and scalable technologies that result in high-quality and reproducible monolayers, making possible the manufacture of thousands of devices. Many advances towards the deposition of a continuous top-contact electrode, without short-circuiting or damaging the monolayer, have also been made in the last years. This potential processability and scalability in the fabrication of large-area molecular electronic devices pave the way towards industrial mass production. In this review, progress in the fabrication of large-area molecular electronic devices in recent years has been presented, with particular emphasis on the techniques used for the fabrication of well-ordered and tightly-packed monolayers, as well as top-contact deposition methodologies. An overview of the current nanofabrication techniques clearly evidences that there is still a long way to go before having a well-defined manufacturing route to get integrated, high yield, robust, stable, scalable, and reproducible molecular devices, produced at a reasonable cost. A multidisciplinary approach with the collaboration of chemists, physicists, and engineers is being made in research groups all over the world to overcome these difficulties and put in the hands of the next generation an emerging technology that is expected to improve people's quality of life.

Funding: This research was funded by Ministerio de Economía y Competitividad from Spain and fondos FEDER in the framework of projects MAT2016-78257-R and PID2019-105881RB-I00. L.H., S.M. and P.C. also acknowledge projects LMP33-18 and E31_20R (Platon research group) funded by Gobierno de Aragón/Fondos FEDER (construyendo Europa desde Aragón).

Conflicts of Interest: The authors declare no conflict of interest.

References

1. Available online: https://www.360marketupdates.com/global-electronic-components-market-14830923 (accessed on 14 August 2020).
2. Lu, W.; Lee, Y.; Murdzek, J.; Gertsch, J.; Vardi, A.; Kong, L.; George, S.; del Alamo, J. First Transistor Demonstration of Thermal Atomic Layer Etching: InGaAs FinFETs with Sub-5 Nm Fin-Width Featuring in Situ ALE-ALD. In Proceedings of the IEEE International Electron Devices Meeting (IEDM), San Francisco, CA, USA, 1–5 December 2018.
3. Waldrop, B.Y.M.M.; Law, O.F.M.S.; Interesting, M. More Than Will Soon Abandon Its Pursuit. *Nature* **2016**, *530*, 145.
4. Available online: https://www.economist.com/technology-quarterly/2016-03-12/after-moores-law (accessed on 14 August 2020).

50. Hsu, L.Y.; Jin, B.Y.; Chen, C.H.; Peng, S.M. Reaction: New Insights into Molecular Electronics. *Chem* **2017**, *3*, 378–379. [CrossRef]

51. Halik, M.; Klauk, H.; Zschieschang, U.; Maisch, S.; Effenberger, F.; Dehm, C.; Schu, M.; Brunnbauer, M.; Stellacci, F. Low-Voltage Organic Transistors with an Amorphous Molecular Gate Dielectric Marcus. *Nature* **2004**, *431*, 963–966. [CrossRef]

52. Aragonès, A.C.; Haworth, N.L.; Darwish, N.; Ciampi, S.; Bloomfield, N.J.; Wallace, G.G.; Diez-Perez, I.; Coote, M.L. Electrostatic Catalysis of a Diels-Alder Reaction. *Nature* **2016**, *531*, 88–91. [CrossRef]

53. Ciampi, S.; Darwish, N.; Aitken, H.M.; Díez-Perez, I.; Coote, M.L. Harnessing Electrostatic Catalysis in Single Molecule, Electrochemical and Chemical Systems: A Rapidly Growing Experimental Tool Box. *Chem. Soc. Rev.* **2018**, *47*, 5146–5164. [CrossRef]

54. Metzger, R.M. Unimolecular Electronics. *Chem. Rev.* **2015**, *115*, 5056–5115. [CrossRef]

55. Pobelov, I.V.; Li, Z.; Wandlowski, T. Electrolyte Gating in Redox-Active Tunneling Junctions—An Electrochemical STM Approach. *J. Am. Chem. Soc.* **2008**, *130*, 16045–16054. [CrossRef]

56. Baghernejad, M.; Manrique, D.Z.; Li, C.; Pope, T.; Zhumaev, U.; Pobelov, I.; Kaliginedi, V.; Huang, C.; Hong, W.; Lambert, C.; et al. Highly-Effective Gating of Single-Molecule Junctions: An Electrochemical Approach. *Chem. Commun.* **2014**, *50*, 15975–15978. [CrossRef]

57. Darwish, N.; Díez-Pérez, I.; Guo, S.; Tao, N.; Gooding, J.J.; Paddon-Row, M.N. Single Molecular Switches: Electrochemical Gating of a Single Anthraquinone-Based Norbornylogous Bridge Molecule. *J. Phys. Chem. C* **2012**, *116*, 21093–21097. [CrossRef]

58. Haga, M.A.; Kobayashi, K.; Terada, K. Fabrication and Functions of Surface Nanomaterials Based on Multilayered or Nanoarrayed Assembly of Metal Complexes. *Coord. Chem. Rev.* **2007**, *251*, 2688–2701. [CrossRef]

59. Rocha, A.R.; García-Suárez, V.M.; Bailey, S.W.; Lambert, C.J.; Ferrer, J.; Sanvito, S. Towards Molecular Spintronics. *Nat. Mater.* **2005**, *4*, 335–339. [CrossRef] [PubMed]

60. Schmaus, S.; Bagrets, A.; Nahas, Y.; Yamada, T.K.; Bork, A.; Bowen, M.; Beaurepaire, E.; Evers, F.; Wulfhekel, W. Giant Magnetoresistance through a Single Molecule. *Nat. Nanotechnol.* **2011**, *6*, 185–189. [CrossRef] [PubMed]

61. Aradhya, S.V.; Venkataraman, L. Single-Molecule Junctions beyond Electronic Transport. *Nat. Nanotechnol.* **2013**, *8*, 399–410. [CrossRef]

62. Ormaza, M.; Abufager, P.; Verlhac, B.; Bachellier, N.; Bocquet, M.L.; Lorente, N.; Limot, L. Controlled Spin Switching in a Metallocene Molecular Junction. *Nat. Commun.* **2017**, *8*, 1974. [CrossRef]

63. Gehring, P.; Thijssen, J.M.; van der Zant, H.S.J. Single-Molecule Quantum-Transport Phenomena in Break Junctions. *Nat. Rev. Phys.* **2019**, *1*, 381–396. [CrossRef]

64. Park, S.; Kang, H.; Yoon, H.J. Structure-Thermopower Relationships in Molecular Thermoelectrics. *J. Mater. Chem. A* **2019**, *7*, 14419–14446. [CrossRef]

65. Perroni, C.A.; Ninno, D.; Cataudella, V. Thermoelectric Efficiency of Molecular Junctions. *J. Phys. Condens. Matter* **2016**, *28*, 373001. [CrossRef]

66. Harzheim, A. Thermoelectricity in Single-Molecule Devices. *Mater. Sci. Technol.* **2018**, *34*, 1275–1286. [CrossRef]

67. Cui, L.; Miao, R.; Jiang, C.; Meyhofer, E.; Reddy, P. Perspective: Thermal and Thermoelectric Transport in Molecular Junctions. *J. Chem. Phys.* **2017**, *146*, 092201. [CrossRef]

68. Cui, L.; Miao, R.; Wang, K.; Thompson, D.; Zotti, L.A.; Cuevas, J.C.; Meyhofer, E.; Reddy, P. Peltier Cooling in Molecular Junctions. *Nat. Nanotechnol.* **2018**, *13*, 122–127. [CrossRef]

69. Gao, W.; Ota, H.; Kiriya, D.; Takei, K.; Javey, A. Flexible Electronics toward Wearable Sensing. *Acc. Chem. Res.* **2019**, *52*, 523–533. [CrossRef] [PubMed]

70. Wang, S.; Oh, J.Y.; Xu, J.; Tran, H.; Bao, Z. Skin-Inspired Electronics: An Emerging Paradigm. *Acc. Chem. Res.* **2018**, *51*, 1033–1045. [CrossRef] [PubMed]

71. Gao, M.; Shih, C.; Chen, W. Advances and Challenges of Green Materials for Electronics and Energy Storage Applications: From Design to End-of-Life Recovery. *J. Mater. Chem. A* **2018**, *6*, 20546–20563. [CrossRef]

72. Bergren, A.J.; Zeer-Wanklyn, L.; Semple, M.; Pekas, N.; Szeto, B.; McCreery, R.L. Musical Molecules: The Molecular Junction as an Active Component in Audio Distortion Circuits. *J. Phys. Condens. Matter* **2016**, *28*, 094011. [CrossRef] [PubMed]

73. Lörtscher, E. Reaction: Technological Aspects of Molecular Electronics. *Chem* **2017**, *3*, 376–377. [CrossRef]

74. Martín, S.; Haiss, W.; Higgins, S.; Cea, P.; López, M.C.; Nichols, R.J. A Comprehensive Study of the Single Molecule Conductance of α,ω-Dicarboxylic Acid-Terminated Alkanes. *J. Phys. Chem. C* **2008**, *112*, 3941–3948. [CrossRef]

75. Xing, Y.; Park, T.H.; Venkatramani, R.; Keinan, S.; Beratan, D.N.; Therien, M.J.; Borguet, E. Optimizing Single-Molecule Conductivity of Conjugated Organic Oligomers with Carbodithioate Linkers. *J. Am. Chem. Soc.* **2010**, *132*, 7946–7956. [CrossRef]

76. Cheng, Z.-L.; Skouta, R.; Vazquez, H.; Widawsky, J.R.; Schneebeli, S.; Chen, W.; Hybertsen, M.S.; Breslow, R.; Venkataraman, L. In Situ Formation of Highly Conducting Covalent Au-C Contacts for Single-Molecule Junctions. *Nat. Nanotechnol.* **2011**, *6*, 353–357. [CrossRef]

77. Weibel, N.; Błaszczyk, A.; von Hänisch, C.; Mayor, M.; Pobelov, I.; Wandlowski, T.; Chen, F.; Tao, N. Redox-Active Catechol-Functionalized Molecular Rods: Suitable Protection Groups and Single-Molecule Transport Investigations. *Eur. J. Org. Chem.* **2008**, *2008*, 136–149. [CrossRef]

78. Arroyo, C.R.; Leary, E.; Castellanos-Gómez, A.; Rubio-Bollinger, G.; González, M.T.; Agraït, N. Influence of Binding Groups on Molecular Junction Formation. *J. Am. Chem. Soc.* **2011**, *133*, 14313–14319. [CrossRef] [PubMed]

79. Pera, G.; Villares, A.; López, M.C.; Cea, P.; Lydon, D.P.; Low, P.J. Preparation and Characterization of Langmuir and Langmuir-Blodgett Films from a Nitrile-Terminated Tolan. *Chem. Mater.* **2007**, *19*, 857–864. [CrossRef]

80. Ferradás, R.R.; Marqués-González, S., Osorio, H.M.; Ferrer, J.; Cea, P.; Milan, D.C.; Vezzoli, A.; Higgins, S.J.; Nichols, R.J.; Low, P.J.; et al. Low Variability of Single-Molecule Conductance Assisted by Bulky Metal-Molecule Contacts. *RSC Adv.* **2016**, *6*, 75111–75121. [CrossRef]

81. Herrer, I.L.; Ismael, A.K.; Milán, D.C.; Vezzoli, A.; Martín, S.; González-Orive, A.; Grace, I.; Lambert, C.; Serrano, J.L.; Nichols, R.J.; et al. Unconventional Single-Molecule Conductance Behavior for a New Heterocyclic Anchoring Group: Pyrazolyl. *J. Phys. Chem. Lett.* **2018**, *9*, 5364–5372. [CrossRef]

82. Valášek, M.; Lindner, M.; Mayor, M. Rigid Multipodal Platforms for Metal Surfaces. *Beilstein J. Nanotechnol.* **2016**, *7*, 374–405. [CrossRef]

83. Osorio, H.M.; Martín, S.; Milan, D.C.; González-Orive, A.; Gluyas, J.B.G.; Higgins, S.J.; Low, P.J.; Nichols, R.J.; Cea, P. Influence of Surface Coverage on the Formation of 4,4'-Bipyridinium (Viologen) Single Molecular Junctions. *J. Mater. Chem. C* **2017**, *5*, 11717–11723. [CrossRef]

84. Walker, A.V. Toward a New World of Molecular Devices: Making Metallic Contacts to Molecules. *J. Vac. Sci. Technol. A* **2013**, *31*, 050816. [CrossRef]

85. Haick, H.; Cahen, D. Making Contact: Connecting Molecules Electrically to the Macroscopic World. *Prog. Surf. Sci.* **2008**, *83*, 217–261. [CrossRef]

86. Song, F.; Wells, J.W.; Handrup, K.; Li, Z.S.; Bao, S.N.; Schulte, K.; Ahola-Tuomi, M.; Mayor, L.C.; Swarbrick, J.C.; Perkins, E.W.; et al. Direct Measurement of Electrical Conductance through a Self-Assembled Molecular Layer. *Nat. Nanotechnol.* **2009**, *4*, 373–376. [CrossRef]

87. Low, P.J. Twists and Turns: Studies of the Complexes and Properties of Bimetallic Complexes Featuring Phenylene Ethynylene and Related Bridging Ligands. *Coord. Chem. Rev* **2013**, *257*, 1507–1532. [CrossRef]

88. Higgins, S.J.; Nichols, R.J.; Martin, S.; Cea, P.; Van Der Zant, H.S.J.; Richter, M.M.; Low, P.J. Looking Ahead: Challenges and Opportunities in Organometallic Chemistry. *Organometallics* **2011**, *30*, 7–12. [CrossRef]

89. Yao, H.; Cui, Y.; Yu, R.; Gao, B.; Zhang, H.; Hou, J. Design, Synthesis, and Photovoltaic Characterization of a Small Molecular Acceptor with an Ultra-Narrow Band Gap. *Angew. Chem. Int. Ed.* **2017**, *56*, 3045–3049. [CrossRef]

90. Vilan, A.; Aswal, D.; Cahen, D. Large-Area, Ensemble Molecular Electronics: Motivation and Challenges. *Chem. Rev.* **2017**, *117*, 4248–4286. [CrossRef]

91. Cea, P.; Ballesteros, L.M.; Martín, S. Nanofabrication Techniques of Highly Organized Monolayers Sandwiched between Two Electrodes for Molecular Electronics. *Nanofabrication* **2014**, *1*, 96–117. [CrossRef]

92. Ramachandran, G.K.; Hopson, T.J.; Rawlett, A.M.; Nagahara, L.A.; Primak, A.; Lindsay, S.M. A Bond-Fluctuation Mechanism for Stochastic Switching in Wired Molecules. *Science* **2003**, *300*, 1413–1416. [CrossRef]

93. Denayer, J.; Delhalle, J.; Mekhalif, Z. Comparative Study of Copper Surface Treatment with Self-Assembled Monolayers of Aliphatic Thiol, Dithiol and Dithiocarboxylic Acid. *J. Electroanal. Chem.* **2009**, *637*, 43–49. [CrossRef]

94. Adaligil, E.; Shon, Y.-S.; Slowinski, K. Effect of Headgroup on Electrical Conductivity of Self-Assembled Monolayers on Mercury: N-Alkanethiols versus n-Alkaneselenols. *Langmuir* **2010**, *26*, 1570–1573. [CrossRef]

95. Beebe, J.M.; Engelkes, V.B.; Miller, L.L.; Frisbie, C.D. Contact Resistance in Metal-Molecule-Metal Junctions Based on Aliphatic SAMs: Effects of Surface Linker and Metal Work Function. *J. Am. Chem. Soc.* **2002**, *124*, 11268–11269. [CrossRef]

96. Tsunoi, A.; Lkhamsuren, G.; Angelo, E.; Mondarte, Q.; Asatyas, S.; Oguchi, M.; Noh, J.; Hayashi, T. Improvement of the Thermal Stability of Self-Assembled Monolayers of Isocyanide Derivatives on Gold. *J. Phys. Chem. C* **2019**, *123*, 13681–13686. [CrossRef]

97. Huo, L.; Du, P.; Zhou, H.; Zhang, K.; Liu, P. Fabrication and Tribological Properties of Self-Assembled Monolayer of n-Alkyltrimethoxysilane on Silicon: Effect of SAM Alkyl Chain Length. *Appl. Surf. Sci.* **2017**, *396*, 865–869. [CrossRef]

98. Wan, X.; Lieberman, I.; Asyuda, A.; Resch, S.; Seim, H.; Zharnikov, M. Thermal Stability of Phosphonic Acid Self-Assembled Monolayers on Alumina Substrates. *J. Phys. Chem. C* **2020**, *124*, 2531–2542. [CrossRef]

99. Adenier, A.; Chehimi, M.M.; Gallardo, I.; Pinson, J.; Vilà, N. Electrochemical Oxidation of Aliphatic Amines and Their Attachment to Carbon and Metal Surfaces. *Langmuir* **2004**, *20*, 8243–8253. [CrossRef] [PubMed]

100. Rittikulsittichai, S.; Park, C.S.; Jamison, A.C.; Rodriguez, D.; Zenasni, O.; Lee, T.R. Bidentate Aromatic Thiols on Gold: New Insight Regarding the Influence of Branching on the Structure, Packing, Wetting, and Stability of Self-Assembled Monolayers on Gold Surfaces. *Langmuir* **2017**, *33*, 4396–4406. [CrossRef]

101. Pathak, A.; Bora, A.; Liao, K.-C.; Schmolke, H.; Jung, A.; Klages, C.-P.; Schwartz, J.; Tornow, M. Disorder-Derived, Strong Tunneling Attenuation in Bis-Phosphonate Monolayers. *J. Phys. Condens. Matter* **2016**, *28*, 094008. [CrossRef] [PubMed]

102. Hopmann, E.; Elezzabi, A.Y. Electrochemical Stability Enhancement of Electrochromic Tungsten Oxide by Self-Assembly of a Phosphonate Protection Layer. *ACS Appl. Mater. Interfaces* **2020**, *12*, 1930–1936. [CrossRef] [PubMed]

103. Camacho-Alanis, F.; Wu, L.; Zangari, G.; Swami, N. Molecular Junctions of ~1 Nm Device Length on Self-Assembled Monolayer Modified n- vs. p-GaAs. *J. Mater. Chem.* **2008**, *18*, 5459–5467. [CrossRef]

104. Villares, A.; Pera, G.; Lydon, D.P.; López, M.C.; Low, P.J.; Cea, P. Mixing Behaviour of a Conjugated Molecular Wire Candidate and an Insulating Fatty Acid within Langmuir-Blodgett Films. *Colloids Surf. A Physicochem. Eng. Asp.* **2009**, *346*, 170–176. [CrossRef]

105. Bartl, J.D.; Gremmo, S.; Stutzmann, M.; Tornow, M.; Cattani-Scholz, A. Modification of Silicon Nitride with Oligo(Ethylene Glycol)-Terminated Organophosphonate Monolayers. *Surf. Sci.* **2020**, *697*, 121599. [CrossRef]

106. Qiu, X.; Ivasyshyn, V.; Qiu, L.; Enache, M.; Dong, J.; Rousseva, S.; Portale, G.; Stöhr, M.; Hummelen, J.C.; Chiechi, R.C. Thiol-Free Self-Assembled Oligoethylene Glycols Enable Robust Air-Stable Molecular Electronics. *Nat. Mater.* **2020**, *19*, 330–337. [CrossRef]

107. Moneo, A.; Gonzalez-Orive, A.; Bock, S.; Fenero, M.; Herrer, L.; Costa-Milan, D.; Lorenzoni, M.; Nichols, R.J.; Cea, P.; Perez-Murano, F.; et al. Towards Molecular Electronic Devices Based on "all-Carbon" Wires. *Nanoscale* **2018**, *10*, 14128–14138. [CrossRef] [PubMed]

108. Gulcur, M.; Moreno-garcía, P.; Zhao, X.; Baghernejad, M. The Synthesis of Functionalised Diaryltetraynes and Their Transport Properties in Single-Molecule Junctions. *Chem. Eur. J.* **2014**, *20*, 4653–4660. [CrossRef]

109. Donhauser, Z.J.; Mantooth, B.A.; Kelly, K.F.; Bumm, L.A.; Monnell, J.D.; Stapleton, J.J.; Price, D.W.; Rawlett, A.M.; Allara, D.L.; Tour, J.M.; et al. Conductance Switching in Single Molecules Through Conformational Changes. *Science* **2001**, *292*, 2303–2307. [CrossRef] [PubMed]

110. Lu, Q.; Liu, K.; Zhang, H.; Du, Z.; Wang, X.; Wang, F. From Tunneling to Hopping: A Comprehensive Investigation of Charge. *ACS Nano* **2009**, *3*, 3861–3868. [CrossRef] [PubMed]

111. Herrer, L.; Sebastian, V.; Martín, S.; González-orive, A.; Pérez-Murano, F.; Low, P.J.; Serrano, J.L.; Santamaría, J.; Cea, P. High Surface Coverage of a Self-Assembled Monolayer by in Situ Synthesis of Palladium Nanodeposits. *Nanoscale* **2017**, *9*, 13281–13291. [CrossRef]

112. Herrer, L.; González-Orive, A.; Marqués-González, S.; Martín, S.; Nichols, R.J.; Serrano, J.L.; Low, P.J.; Cea, P. Electrically Transmissive Alkyne-Anchored Monolayers on Gold. *Nanoscale* **2019**, *11*, 7976–7985. [CrossRef]

113. Pera, G.; Martín, S.; Ballesteros, L.M.; Hope, A.J.; Low, P.J.; Nichols, R.J.; Cea, P. Metal-Molecule-Metal Junctions in Langmuir-Blodgett Films Using a New Linker: Trimethylsilane. *Chem. Eur. J.* **2010**, *16*, 13398–13405. [CrossRef]

114. Ballesteros, L.M.; Martín, S.; Pera, G.; Schauer, P.A.; Kay, N.J.; López, M.C.; Low, P.J.; Nichols, R.J.; Cea, P. Directionally Oriented LB Films of an OPE Derivative: Assembly, Characterization, and Electrical Properties. *Langmuir* **2011**, *27*, 3600–3610. [CrossRef]

115. Martín, S.; Ballesteros, L.M.; González-Orive, A.; Oliva, H.; Marqués-González, S.; Lorenzoni, M.; Nichols, R.J.; Pérez-Murano, F.; Low, P.J.; Cea, P. Towards a Metallic Top Contact Electrode in Molecular Electronic Devices Exhibiting a Large Surface Coverage by Photoreduction of Silver Cations. *J. Mater. Chem. C* **2016**, *4*, 9036–9043. [CrossRef]

116. Villares, A.; Lydon, D.P.; Porrés, L.; Beeby, A.; Low, P.J.; Cea, P.; Royo, F.M. Preparation of Ordered Films Containing a Phenylene Ethynylene Oligomer by the Langmuir-Blodgett Technique. *J. Phys. Chem. B* **2007**, *111*, 7201–7209. [CrossRef]

117. Sangiao, S.; Martín, S.; González-Orive, A.; Magén, C.; Low, P.J.; De Teresa, J.M.; Cea, P. All-Carbon Electrode Molecular Electronic Devices Based on Langmuir-Blodgett Monolayers. *Small* **2017**, *13*, 1603207. [CrossRef] [PubMed]

118. Villares, A.; Pera, G.; Martin, S.; Nichols, R.J.; Lydon, D.P.; Applegarth, L.; Beeby, A.; Low, P.J.; Cea, P. Fabrication, Characterization, and Electrical Properties of Langmuir-Blodgett Films of an Acid Terminated Phenylene-Ethynylene Oligomer. *Chem. Mater.* **2010**, *22*, 2041–2049. [CrossRef]

119. Osorio, H.M.; Cea, P.; Ballesteros, L.M.; Gascón, I.; Marqués-González, S.; Nichols, R.J.; Pérez-Murano, F.; Low, P.J.; Martín, S. Preparation of Nascent Molecular Electronic Devices from Gold Nanoparticles and Terminal Alkyne Functionalised Monolayer Films. *J. Mater. Chem. C* **2014**, *2*, 7348–7355. [CrossRef]

120. Ballesteros, L.M.; Martín, S.; Momblona, C.; Marqués-González, S.; López, M.C.; Nichols, R.J.; Low, P.J.; Cea, P. Acetylene Used as a New Linker for Molecular Junctions in Phenylene—Ethynylene Oligomer Langmuir–Blodgett Films. *J. Phys. Chem. C* **2012**, *116*, 9142–9150. [CrossRef]

121. Villares, A.; Lydon, D.P.; Robinson, B.J.; Ashwell, G.J.; Royo, F.M.; Low, P.J.; Cea, P. Langmuir-Blodgett Films Incorporating Molecular Wire Candidates of Ester-Substituted Oligo(Phenylene-Ethynylene) Derivatives. *Surf. Sci.* **2008**, *602*, 3683–3687. [CrossRef]

122. Casalini, S.; Berto, M.; Leonardi, F.; Operamolla, A.; Bortolotti, C.A.; Borsari, M.; Sun, W.; Di Felice, R.; Corni, S.; Albonetti, C.; et al. Self-Assembly of Mono- and Bidentate Oligoarylene Thiols onto Polycrystalline Au. *Langmuir* **2013**, *29*, 13198–13208. [CrossRef]

123. Von Wrochem, F.; Gao, D.; Scholz, F.; Nothofer, H.-G.; Nelles, G.; Wessels, J.M. Efficient Electronic Coupling and Improved Stability with Dithiocarbamate-Based Molecular Junctions. *Nat. Nanotechnol.* **2010**, *5*, 618–624. [CrossRef]

124. Pla-Vilanova, P.; Aragonès, A.C.; Ciampi, S.; Sanz, F.; Darwish, N.; Díez-Pérez, I. The Spontaneous Formation of Single-Molecule Junctions via Terminal Alkynes. *Nanotechnology* **2015**, *26*, 381001. [CrossRef]

125. Osorio, H.M.; Martín, S.; López, M.C.; Marqués-González, S.; Higgins, S.J.; Nichols, R.J.; Low, P.J.; Cea, P. Electrical Characterization of Single Molecule and Langmuir-Blodgett Monomolecular Films of a Pyridine Terminated Oligo(Phenylene-Ethynylene) Derivative. *Beilstein J. Nanotechnol.* **2015**, *6*, 1145–1157. [CrossRef]

126. Escorihuela, E.; Cea, P.; Naghibi, S.; Osorio, H.M.; Martin, S.; Milan, D.C.; Nichols, R.J.; Low, P.J. Towards the Design of Effective Multipodal Contacts for Use in the Construction of Langmuir—Blodgett Films and Molecular Junctions. *J. Mater. Chem. A* **2020**, *8*, 672–682. [CrossRef]

127. Herrer, L.; Ismael, A.; Martín, S.; Milan, D.C.C.; Serrano, J.L.L.; Nichols, R.J.J.; Lambert, C.; Cea, P. Single Molecule vs. Large Area Design of Molecular Electronic Devices Incorporating an Efficient 2-Aminepyridine Double Anchoring Group. *Nanoscale* **2019**, *11*, 15871–15880. [CrossRef] [PubMed]

128. Baranton, S.; Bélanger, D. In Situ Generation of Diazonium Cations in Organic Electrolyte for Electrochemical Modification of Electrode Surface. *Electrochim. Acta* **2008**, *53*, 6961–6967. [CrossRef]

129. Liu, W.; Tilley, T.D. Sterically Controlled Functionalization of Carbon Surfaces With-$C_6H_4CH_2X$ (X = OSO_2Me or N_3) Groups for Surface Attachment of Redox-Active Molecules. *Langmuir* **2015**, *31*, 1189–1195. [CrossRef] [PubMed]

130. Combellas, C.; Kanoufi, F.; Pinson, J.; Podvorica, F.I. Indirect Electrografting of Aryl Iodides. *Electrochem. Commun.* **2019**, *98*, 119–123. [CrossRef]

131. Ditzler, L.R.; Karunatilaka, C.; Donuru, V.R.; Liu, H.Y.; Tivanski, A.V. Electromechanical Properties of Self-Assembled Monolayers of Tetrathiafulvalene Derivatives Studied by Conducting Probe Atomic Force Microscopy. *J. Phys. Chem. C* **2010**, *114*, 4429–4435. [CrossRef]

132. Yokoyama, T.; Kawasaki, M.; Asari, T.; Ohno, S.; Tanaka, M.; Yoshimoto, Y. Adsorption and Self-Assembled Structures of Sexithiophene on the Si(111)- 3×3 -Ag Surface. *J. Chem. Phys.* **2015**, *142*, 204701. [CrossRef]

133. Parry, A.V.S.; Lu, K.; Tate, D.J.; Urasinska-Wojcik, B.; Caras-Quintero, D.; Majewski, L.A.; Turner, M.L. Trichlorosilanes as Anchoring Groups for Phenylene-Thiophene Molecular Monolayer Field Effect Transistors. *Adv. Funct. Mater.* **2014**, *24*, 6677–6683. [CrossRef]

134. Sebechlebska, T.; Sebera, J.; Kolivoska, V.; Lindner, M.; Gasior, J.; Mészáros, G.; Valásek, M.; Mayor, M.; Hromadová, M. Investigation of the Geometrical Arrangement and Single Molecule Charge Transport in Self-Assembled Monolayers of Molecular Towers Based on Tetraphenylmethane Tripod. *Electrochim. Acta* **2017**, *258*, 1191–1200. [CrossRef]

135. Valášek, M.; Mayor, M.; Valáseck, M.; Mayor, M. Spatial and Lateral Control of Functionality by Rigid Molecular Platforms. *Chem. Eur. J.* **2017**, *23*, 13538–13548. [CrossRef]

136. Ye, Q.; Wang, H.; Yu, B.; Zhou, F. Self-Assembly of Catecholic Ferrocene and Electrochemical Behavior of Its Monolayer. *RSC Adv.* **2015**, *5*, 60090–60095. [CrossRef]

137. Ballesteros, L.M.; Martín, S.; Cortés, J.; Marqués-González, S.; Pérez-Murano, F.; Nichols, R.J.; Low, P.J.; Cea, P. From an Organometallic Monolayer to an Organic Monolayer Covered by Metal Nanoislands: A Simple Thermal Protocol for the Fabrication of the Top Contact Electrode in Molecular Electronic Devices. *Adv. Mater. Interfaces* **2014**, *1*, 1400128. [CrossRef]

138. Ezquerra, R.; Eaves, S.G.; Bock, S.; Skelton, B.W.; Pérez-Murano, F.; Cea, P.; Martín, S.; Low, P.J. New Routes to Organometallic Molecular Junctions: Via a Simple Thermal Processing Protocol. *J. Mater. Chem. C* **2019**, *7*, 6630–6640. [CrossRef]

139. Tefashe, U.M.; Van Nguyen, Q.; Lafolet, F.; Lacroix, J.C.; McCreery, R.L. Robust Bipolar Light Emission and Charge Transport in Symmetric Molecular Junctions. *J. Am. Chem. Soc.* **2017**, *139*, 7436–7439. [CrossRef] [PubMed]

140. Yasseri, A.A.; Syomin, D.; Malinovskii, V.L.; Loewe, R.S.; Lindsey, J.S.; Zaera, F.; Bocian, D.F. Characterization of Self-Assembled Monolayers of Porphyrins Bearing Multiple Thiol-Derivatized Rigid-Rod Tethers. *J. Am. Chem. Soc.* **2004**, *126*, 11944–11953. [CrossRef]

141. Otte, F.L.; Lemke, S.; Schütt, C.; Krekiehn, N.R.; Jung, U.; Magnussen, O.M.; Herges, R. Ordered Monolayers of Free-Standing Porphyrins on Gold. *J. Am. Chem. Soc.* **2014**, *136*, 11248–11251. [CrossRef]

142. He, W.-L.; Fang, F.; Ma, D.-M.; Chen, M.; Qian, D.-J.; Liu, M. Palladium-Directed Self-Assembly of Multi-Titanium(IV)-Porphyrin Arrays on the Substrate Surface as Sensitive Ultrathin Films for Hydrogen Peroxide Sensing, Photocurrent Generation, and Photochromism of Viologen. *Appl. Surf. Sci.* **2018**, *427*, 1003–1010. [CrossRef]

143. Kim, Y.-S.; Fournier, S.; Lau-Truong, S.; Decorse, P.; Devillers, C.H.; Lucas, D.; Harris, K.D.; Limoges, B.; Balland, V. Introducing Molecular Functionalities within High Surface Area Nanostructured ITO Electrodes through Diazonium Electrografting. *ChemElectroChem* **2018**, *5*, 1625–1630. [CrossRef]

144. Zhang, Z.; Yoshida, N.; Imae, T.; Xue, Q.; Bai, M.; Jiang, J.; Liu, Z. A Self-Assembled Monolayer of an Alkanoic Acid-Derivatized Porphyrin on Gold Surface: A Structural Investigation by Surface Plasmon Resonance, Ultraviolet–Visible, and Infrared Spectroscopies. *J. Colloid Interface Sci.* **2001**, *243*, 382–387. [CrossRef]

145. Bodik, M.; Maxian, O.; Hagara, J.; Nadazdy, P.; Jergel, M.; Majkova, E.; Siffalovic, P. Langmuir–Scheaffer Technique as a Method for Controlled Alignment of 1D Materials. *Langmuir* **2020**, *36*, 4540–4547. [CrossRef]

146. Kh Dzhanabekova, R.; Seliverstova, E.V.; Zh Zhumabekov, A.; Kh Ibrayev, N. Fabricating and Examining of Langmuir Films of Reduced Graphene Oxide. *Russ. J. Phys. Chem. A* **2019**, *93*, 284–289. [CrossRef]

147. Daraghma, S.M.A.; Talebi, S.; Zhijian, C.; Periasamy, V. Method of Assembling Pure Langmuir–Blodgett DNA Films Using TBE Buffer as the Subphase. *Appl. Phys. A Mater. Sci. Process.* **2019**, *125*, 593. [CrossRef]

148. Fereiro, J.A.; Yu, X.; Pecht, I.; Sheves, M.; Cuevas, J.C.; Cahen, D. Tunneling Explains Efficient Electron Transport via Protein Junctions. *Proc. Natl. Acad. Sci. USA* **2018**, *115*, E4577–E4583. [CrossRef] [PubMed]

149. Fereiro, J.A.; Kayser, B.; Romero-Muñiz, C.; Vilan, A.; Dolgikh, D.A.; Chertkova, R.V.; Cuevas, J.C.; Zotti, L.A.; Pecht, I.; Sheves, M.; et al. A Solid-State Protein Junction Serves as a Bias-Induced Current Switch. *Angew. Chem. Int. Ed.* **2019**, *58*, 11852–11859. [CrossRef] [PubMed]

150. Whitesides, G.; Grzybowski, B. Self-Assembly at All Scales. *Science* **2002**, *295*, 2418–2421. [CrossRef]

151. Casalini, S.; Bortolotti, C.A.; Leonardi, F.; Biscarini, F. Self-Assembled Monolayers in Organic Electronics. *Chem. Soc. Rev.* **2017**, *46*, 40–71. [CrossRef]

152. Desbief, S.; Patrone, L.; Goguenheim, D.; Guérin, D.; Vuillaume, D. Impact of Chain Length, Temperature, and Humidity on the Growth of Long Alkyltrichlorosilane Self-Assembled Monolayers. *Phys. Chem. Chem. Phys.* **2011**, *13*, 2870–2879. [CrossRef]

153. Akkerman, H.B.; Kronemeijer, A.J.; Van Hal, P.A.; De Leeuw, D.M.; Blom, P.W.M.; De Boer, B. Self-Assembled-Monolayer Formation of Long Alkanedithiols in Molecular Junctions. *Small* **2008**, *4*, 100–104. [CrossRef]

154. Bigelow, W.C.; Pickett, D.L.; Zisman, W.A. Oleophobic Monolayers: I. Films Adsorbed from Solution in Non-Polar Liquids. *J. Colloid Sci.* **1946**, *1*, 513–538. [CrossRef]

155. Sagiv, J. Organized Monolayers by Adsorption. 1. Formation and Structure of Oleophobic Mixed Monolayers on Solid Surfaces. *J. Am. Chem. Soc.* **1980**, *102*, 92–98. [CrossRef]

156. Nuzzo, R.G.; Allara, D.L. Adsorption of Bifunctional Organic Disulfides on Gold Surfaces. *J. Am. Chem. Soc.* **1983**, *105*, 4481–4483. [CrossRef]

157. Ashwell, G.J.; Williams, A.T.; Barnes, S.A.; Chappell, S.L.; Phillips, L.J.; Robinson, B.J.; Urasinska-Wojcik, B.; Wierzchowiec, P.; Gentle, I.R.; Wood, B.J. Self-Assembly of Amino-Thiols via Gold-Nitrogen Links and Consequence for in Situ Elongation of Molecular Wires on Surface-Modified Electrodes. *J. Phys. Chem. C* **2011**, *115*, 4200–4208. [CrossRef]

158. Sims, R.A.; Harmer, S.L.; Quinton, J.S. The Role of Physisorption and Chemisorption in the Oscillatory Adsorption of Organosilanes on Aluminium Oxide. *Polymers* **2019**, *11*, 410. [CrossRef] [PubMed]

159. Salvarezza, R.C.; Carro, P. The Electrochemical Stability of Thiols on Gold Surfaces. *J. Electroanal. Chem.* **2018**, *819*, 234–239. [CrossRef]

160. Xie, Z.; Bâldea, I.; Frisbie, C.D. Determination of Energy-Level Alignment in Molecular Tunnel Junctions by Transport and Spectroscopy: Self-Consistency for the Case of Oligophenylene Thiols and Dithiols on Ag, Au, and Pt Electrodes. *J. Am. Chem. Soc.* **2019**, *141*, 3670–3681. [CrossRef] [PubMed]

161. Vericat, C.; Vela, M.E.; Benitez, G.; Carro, P.; Salvarezza, R.C. Self-Assembled Monolayers of Thiols and Dithiols on Gold: New Challenges for a Well-Known System. *Chem. Soc. Rev.* **2010**, *39*, 1805–1834. [CrossRef] [PubMed]

162. Muglali, M.I.; Erbe, A.; Chen, Y.; Barth, C.; Koelsch, P.; Rohwerder, M. Modulation of Electrochemical Hydrogen Evolution Rate by Araliphatic Thiol Monolayers on Gold. *Electrochim. Acta* **2013**, *90*, 17–26. [CrossRef]

163. Love, J.C.; Estroff, L.A.; Kriebel, J.K.; Nuzzo, R.G.; Whitesides, G.M. Self-Assembled Monolayers of Thiolates on Metals as a Form of Nanotechnology. *Chem. Rev.* **2005**, *105*, 1103–1169. [CrossRef]

164. Wang, Y.; Chi, Q.; Hush, N.S.; Reimers, J.R.; Zhang, J.; Ulstrup, J. Scanning Tunneling Microscopic Observation of Adatom-Mediated Motifs on Gold- Thiol Self-Assembled Monolayers at High Coverage. *J. Phys. Chem. C* **2009**, *113*, 19601–19608. [CrossRef]

165. Kim, C.M.; Bechhoefer, J. Conductive Probe AFM Study of Pt-Thiol and Au-Thiol Contacts in Metal-Molecule-Metal Systems. *J. Chem. Phys.* **2013**, *138*, 014707. [CrossRef]

166. Casalini, S.; Leonardi, F.; Bortolotti, C.A.; Operamolla, A.; Omar, O.H.; Paltrinieri, L.; Albonetti, C.; Farinola, G.M.; Biscarini, F. Mono/Bidentate Thiol Oligoarylene-Based Self-Assembled Monolayers (SAMs) for Interface Engineering. *J. Mater. Chem.* **2012**, *22*, 12155–12163. [CrossRef]

167. Gronbeck, H.; Curioni, A.; Andreoni, W. Thiols and Disulfides on the Au (111) Surface: The Headgroup - Gold Interaction. *J. Am. Chem. Soc.* **2000**, *122*, 3839–3842. [CrossRef]

168. Bürgi, T. Properties of the Gold–Sulphur Interface: From Self-Assembled Monolayers to Clusters. *Nanoscale* **2015**, *7*, 15553–15567. [CrossRef] [PubMed]

169. Zhang, Y.; Qiu, X.; Gordiichuk, P.; Soni, S.; Krijger, T.L.; Herrmann, A.; Chiechi, R.C. Mechanically and Electrically Robust Self-Assembled Monolayers for Large-Area Tunneling Junctions. *J. Phys. Chem. C* **2017**, *121*, 14920–14928. [CrossRef] [PubMed]

170. Lindner, M.; Valášek, M.; Homberg, J.; Edelmann, K.; Gerhard, L.; Wulfhekel, W.; Fuhr, O.; Wächter, T.; Zharnikov, M.; Kolivoška, V.; et al. Importance of the Anchor Group Position (Para versus Meta) in Tetraphenylmethane Tripods: Synthesis and Self-Assembly Features. *Chem. Eur. J.* **2016**, *22*, 13218–13235. [CrossRef] [PubMed]

171. Carro, P.; Torrelles, X.; Salvarezza, R.C. A Novel Model for the $(\sqrt{3}\times\sqrt{3})R30°$ Alkanethiolate–Au(111) Phase Based on Alkanethiolate–Au Adatom Complexes. *Phys. Chem. Chem. Phys.* **2014**, *16*, 19017–19023. [CrossRef]

172. Xue, Y.; Li, X.; Li, H.; Zhang, W. Quantifying Thiol-Gold Interactions towards the Efficient Strength Control. *Nat. Commun.* **2014**, *5*, 4348. [CrossRef]

173. Miller, C.; Cuendet, P.; Graetzel, M. Adsorbed ω-Hydroxy Thiol Monolayers on Gold Electrodes: Evidence for Electron Tunneling to Redox Species in Solution. *J. Phys. Chem.* **1991**, *95*, 877–886. [CrossRef]

174. Porter, M.D.; Bright, T.B.; Allara, D.L.; Chidsey, C.E. Spontaneously Organized Molecular Assemblies. 4. Structural Characterization of n-Alkyl Thiol Monolayers on Gold by Optical Ellipsometry, Infrared Spectroscopy, and Electrochemistry. *J. Am. Chem. Soc.* **1987**, *109*, 3559–3568. [CrossRef]

175. Felgenhauer, T.; Rong, H.; Buck, M. Electrochemical and Exchange Studies of Self-Assembled Monolayers of Biphenyl Based Thiols on Gold. *J. Electroanal. Chem.* **2003**, *551*, 309–319. [CrossRef]

176. Reed, M.A.; Zhou, C.; Muller, C.J.; Burgin, T.P.; Tour, J.M. Conductance of a Molecular Junction. *Science* **1997**, *278*, 252–254. [CrossRef]

177. Dorogi, M.; Gomez, J.; Osifchin, R.; Andres, R.P.; Reifenberger, R. Room-Temperature Coulomb Blockade from a Self-Assembled Molecular Nanostructure. *Phys. Rev. B* **1995**, *52*, 9071–9077. [CrossRef] [PubMed]

178. Gittins, D.I.; Bethell, D.; Schiffrin, D.J.; Nichols, R.J. A Nanometre-Scale Electronic Switch Consisting of a Metal Cluster and Redox-Addressable Groups. *Nature* **2000**, *408*, 67–69. [CrossRef] [PubMed]

179. Cui, X.D.; Primak, A.; Zarate, X.; Tomfohr, J.; Sankey, O.F.; Moore, A.L.; Moore, T.A.; Gust, D.; Harris, G.; Lindsay, S.M. Reproducible Measurement of Single-Molecule Conductivity. *Science* **2001**, *294*, 571–574. [CrossRef] [PubMed]

180. Stapleton, J.J.; Harder, P.; Daniel, T.A.; Reinard, M.D.; Yao, Y.; Price, D.W.; Tour, J.M.; Allara, D.L. Self-Assembled Oligo(Phenylene-Ethynylene) Molecular Electronic Switch Monolayers on Gold: Structures and Chemical Stability. *Langmuir* **2003**, *19*, 8245–8255. [CrossRef]

181. Inkpen, M.S.; Liu, Z.; Li, H.; Campos, L.M.; Neaton, J.B.; Venkataraman, L. Non-Chemisorbed Gold–Sulfur Binding Prevails in Self-Assembled Monolayers. *Nat. Chem.* **2019**, *11*, 351–358. [CrossRef]

182. Kang, H.; Noh, J.; Ganbold, E.O.; Uuriintuya, D.; Gong, M.S.; Oh, J.J.; Joo, S.W. Adsorption Changes of Cyclohexyl Isothiocyanate on Gold Surfaces. *J. Colloid Interface Sci.* **2009**, *336*, 648–653. [CrossRef]

183. Katsonis, N.; Marchenko, A.; Taillemite, S.; Fichou, D.; Chouraqui, G.; Aubert, C.; Malacria, M. A Molecular Approach to Self-Assembly of Trimethylsilylacetylene Derivatives on Gold. *Chem. Eur. J.* **2003**, *9*, 2574–2581. [CrossRef]

184. Katsonis, N.; Marchenko, A.; Fichou, D.; Barrett, N. Investigation on the Nature of the Chemical Link between Acetylenic Organosilane Self-Assembled Monolayers and Au (111) by Means of Synchrotron Radiation Photoelectron Spectroscopy and Scanning Tunneling Microscopy. *Surf. Sci.* **2008**, *602*, 9–16. [CrossRef]

185. Bejarano, F.; Olavarria-Contreras, I.J.; Droghetti, A.; Rungger, I.; Rudnev, A.; Gutiérrez, D.; Mas-Torrent, M.; Veciana, J.; van der Zant, H.S.J.; Rovira, C.; et al. Robust Organic Radical Molecular Junctions Using Acetylene Terminated Groups for C–Au Bond Formation. *J. Am. Chem. Soc.* **2018**, *140*, 1691–1696. [CrossRef]

186. Eder, K.; Felfer, P.J.; Gault, B.; Ceguerra, A.V.; La Fontaine, A.; Masters, A.F.; Maschmeyer, T.; Cairney, J.M. A New Approach to Understand the Adsorption of Thiophene on Different Surfaces: An Atom Probe Investigation of Self-Assembled Monolayers. *Langmuir* **2017**, *33*, 9573–9581. [CrossRef]

187. Gu, C.; Zhang, J.L.; Sun, S.; Lian, X.; Ma, Z.; Mao, H.; Guo, L.; Wang, Y.; Chen, W. Molecular-Scale Investigation of the Thermal and Chemical Stability of Monolayer PTCDA on Cu(111) and Cu(110). *ACS Appl. Mater. Interfaces* **2020**, *12*, 22327–22334. [CrossRef] [PubMed]

188. Martin, C.A.; Ding, D.; Sørensen, J.K.; Bjørnholm, T.; van Ruitenbeek, J.M.; van der Zant, H.S.J. Fullerene-Based Anchoring Groups for Molecular Electronics. *J. Am. Chem. Soc.* **2008**, *130*, 13198–13199. [CrossRef] [PubMed]

189. Del Carmen Gimenez-Lopez, M.; Räisänen, M.T.; Chamberlain, T.W.; Weber, U.; Lebedeva, M.; Rance, G.A.; Andrew, G.; Briggs, D.; Pettifor, D.; Burlakov, V.; et al. Functionalized Fullerenes in Self-Assembled Monolayers. *Langmuir* **2011**, *27*, 10977–10985. [CrossRef] [PubMed]

190. Chinwangso, P.; Jamison, A.C.; Lee, T.R. Multidentate Adsorbates for Self-Assembled Monolayer Films. *Acc. Chem. Res.* **2011**, *44*, 511–519. [CrossRef] [PubMed]

191. Gao, D.; Scholz, F.; Nothofer, H.-G.; Ford, W.E.; Scherf, U.; Wessels, J.M.; Yasuda, A.; von Wrochem, F. Fabrication of Asymmetric Molecular Junctions by the Oriented Assembly of Dithiocarbamate Rectifiers. *J. Am. Chem. Soc.* **2011**, *133*, 5921–5930. [CrossRef]

192. Terada, K.I.; Nakamura, H.; Kanaizuka, K.; Haga, M.A.; Asai, Y.; Ishida, T. Long-Range Electron Transport of Ruthenium-Centered Multilayer Films via a Stepping-Stone Mechanism. *ACS Nano* **2012**, *6*, 1988–1999. [CrossRef]

193. Weidner, T.; Ballav, N.; Siemeling, U.; Troegel, D.; Walter, T.; Tacke, R.; Castner, D.G.; Zharnikov, M. Tripodal Binding Units for Self-Assembled Monolayers on Gold: A Comparison of Thiol and Thioether Headgroups. *J. Phys. Chem. C* **2009**, *113*, 19609–19617. [CrossRef]

194. O'Driscoll, L.J.; Wang, X.; Jay, M.; Batsanov, A.S.; Sadeghi, H.; Lambert, C.J.; Robinson, B.J.; Bryce, M.R. Carbazole-Based Tetrapodal Anchor Groups for Gold Surfaces: Synthesis and Conductance Properties. *Angew. Chem. Int. Ed.* **2020**, *59*, 882–889. [CrossRef]

195. Darwish, N.; Aragonès, A.C.; Darwish, T.; Ciampi, S.; Díez-Pérez, I. Multi-Responsive Photo- and Chemo-Electrical Single-Molecule Switches. *Nano Lett.* **2014**, *14*, 7064–7070. [CrossRef]

196. Ie, Y.; Tanaka, K.; Tashiro, A.; Lee, S.K.; Testai, H.R.; Yamada, R.; Tada, H.; Aso, Y. Thiophene-Based Tripodal Anchor Units for Hole Transport in Single-Molecule Junctions with Gold Electrodes. *J. Phys. Chem. Lett.* **2015**, *6*, 3754–3759. [CrossRef]

197. Ie, Y.; Hirose, T.; Nakamura, H.; Kiguchi, M.; Takagi, N.; Kawai, M.; Aso, Y. Nature of Electron Transport by Pyridine-Based Tripodal Anchors: Potential for Robust and Conductive Single-Molecule Junctions with Gold Electrodes. *J. Am. Chem. Soc.* **2011**, *133*, 3014–3022. [CrossRef] [PubMed]

198. Gerhard, L.; Edelmann, K.; Homberg, J.; Valášek, M.; Bahoosh, S.G.; Lukas, M.; Pauly, F.; Mayor, M.; Wulfhekel, W. An Electrically Actuated Molecular Toggle Switch. *Nat. Commun.* **2017**, *8*, 14672. [CrossRef] [PubMed]

199. Kong, G.D.; Byeon, S.E.; Park, S.; Song, H.; Kim, S.Y.; Yoon, H.J. Mixed Molecular Electronics: Tunneling Behaviors and Applications of Mixed Self-Assembled Monolayers. *Adv. Electron. Mater.* **2020**, *6*, 1901157. [CrossRef]

200. Saegusa, T.; Sakai, H.; Nagashima, H.; Kobori, Y.; Tkachenko, N.V.; Hasobe, T. Controlled Orientations of Neighboring Tetracene Units by Mixed Self-Assembled Monolayers on Gold Nanoclusters for High-Yield and Long-Lived Triplet Excited States through Singlet Fission. *J. Am. Chem. Soc.* **2019**, *141*, 14720–14727. [CrossRef] [PubMed]

201. Huang, X.; Li, T. Recent Progress in the Development of Molecular-Scale Electronics Based on Photoswitchable Molecules. *J. Mater. Chem. C* **2020**, *8*, 821–848. [CrossRef]

202. Rohwerder, M.; de Weldige, K.; Stratmann, M. Potential Dependence of the Kinetics of Thiol Self-Organization on Au(111). *J. Solid State Electrochem.* **1998**, *2*, 88–93. [CrossRef]

203. Pillai, R.G.; Braun, M.D.; Freund, M.S. Electrochemically Assisted Self-Assembly of Alkylthiosulfates and Alkanethiols on Gold: The Role of Gold Oxide Formation and Corrosion. *Langmuir* **2010**, *26*, 269–276. [CrossRef]

204. Sahli, R.; Fave, C.; Raouafi, N.; Boujlel, K.; Schöllhorn, B.; Limoges, B. Switching On/Off the Chemisorption of Thioctic-Based Self-Assembled Monolayers on Gold by Applying a Moderate Cathodic/Anodic Potential. *Langmuir* **2013**, *29*, 5360–5368. [CrossRef]

205. Metoki, N.; Liu, L.; Beilis, E.; Eliaz, N.; Mandler, D. Preparation and Characterization of Alkylphosphonic Acid Self-Assembled Monolayers on Titanium Alloy by Chemisorption and Electrochemical Deposition. *Langmuir* **2014**, *30*, 6791–6799. [CrossRef]

206. Fioravanti, G.; Lugli, F.; Gentili, D.; Mucciante, V.; Leonardi, F.; Pasquali, L.; Liscio, A.; Murgia, M.; Zerbetto, F.; Cavallini, M. Electrochemical Fabrication of Surface Chemical Gradients in Thiol Self-Assembled Monolayers with Tailored Work-Functions. *Langmuir* **2014**, *30*, 11591–11598. [CrossRef]

207. Kwok, S.C.T.; Ciucci, F.; Yuen, M.M.F. Chemisorption Threshold of Thiol-Based Monolayer on Copper: Effect of Electric Potential and Elevated Temperature. *Electrochim. Acta* **2016**, *198*, 185–194. [CrossRef]

208. Zaba, T.; Noworolska, A.; Bowers, C.M.; Breiten, B.; Whitesides, G.M.; Cyganik, P. Formation of Highly Ordered Self-Assembled Monolayers of Alkynes on Au(111) Substrate. *J. Am. Chem. Soc.* **2014**, *136*, 11918–11921. [CrossRef]

209. Widawsky, J.R.; Chen, W.; Vázquez, H.; Kim, T.; Breslow, R.; Hybertsen, M.S.; Venkataraman, L. Length-Dependent Thermopower of Highly Conducting Au–C Bonded Single Molecule Junctions. *Nano Lett.* **2013**, *13*, 2889–2894. [CrossRef] [PubMed]

210. Chen, W.; Widawsky, J.R.; Vázquez, H.; Schneebeli, S.T.; Hybertsen, M.S.; Breslow, R.; Venkataraman, L. Highly Conducting π-Conjugated Molecular Junctions Covalently Bonded to Gold Electrodes. *J. Am. Chem. Soc.* **2011**, *133*, 17160–17163. [CrossRef] [PubMed]

211. Hong, W.; Li, H.; Liu, S.-X.; Fu, Y.; Li, J.; Kaliginedi, V.; Decurtins, S.; Wandlowski, T. Trimethylsilyl-Terminated Oligo(Phenylene Ethynylene)s: An Approach to Single-Molecule Junctions with Covalent Au-C σ-Bonds. *J. Am. Chem. Soc.* **2012**, *134*, 19425–19431. [CrossRef] [PubMed]

212. Fu, Y.; Chen, S.; Kuzume, A.; Rudnev, A.; Huang, C.; Kaliginedi, V.; Baghernejad, M.; Hong, W.; Wandlowski, T.; Decurtins, S.; et al. Exploitation of Desilylation Chemistry in Tailor-Made Functionalization on Diverse Surfaces. *Nat. Commun.* **2015**, *6*, 6403. [CrossRef]

213. Crudden, C.M.; Horton, J.H.; Ebralidze, I.I.; Zenkina, O.V.; McLean, A.B.; Drevniok, B.; She, Z.; Kraatz, H.-B.; Mosey, N.J.; Seki, T.; et al. Ultra Stable Self-Assembled Monolayers of N-Heterocyclic Carbenes on Gold. *Nat. Chem.* **2014**, *6*, 409–414. [CrossRef]

214. Zhao, J.; Davis, J.J. Force Dependent Metalloprotein Conductance by Conducting Atomic Force Microscopy. *Nanotechnology* **2003**, *14*, 1023–1028. [CrossRef]

215. Frascerra, V.; Calabi, F.; Maruccio, G.; Pompa, P.P.; Cingolani, R.; Rinaldi, R. Resonant Electron Tunneling through Azurin in Air and Liquid by Scanning Tunneling Microscopy. *IEEE Trans. Nanotechnol.* **2005**, *4*, 637–640. [CrossRef]

216. Artés, J.M.; López-Martínez, M.; Giraudet, A.; Díez-Pérez, I.; Sanz, F.; Gorostiza, P. Current–Voltage Characteristics and Transition Voltage Spectroscopy of Individual Redox Proteins. *J. Am. Chem. Soc.* **2012**, *134*, 20218–20221. [CrossRef]

217. Bâldea, I. Important Insight into Electron Transfer in Single-Molecule Junctions Based on Redox Metalloproteins from Transition Voltage Spectroscopy. *J. Phys. Chem. C* **2013**, *117*, 25798–25804. [CrossRef]

218. Ruiz, M.P.; Aragonès, A.C.; Camarero, N.; Vilhena, J.G.; Ortega, M.; Zotti, L.A.; Pérez, R.; Cuevas, J.C.; Gorostiza, P.; Díez-Pérez, I. Bioengineering a Single-Protein Junction. *J. Am. Chem. Soc.* **2017**, *139*, 15337–15346. [CrossRef] [PubMed]

219. Sagara, T.; Nakajima, S.; Akutsu, H.; Niki, K.; Wilson, G.S. Heterogeneous Electron-Transfer Rate Measurements of Cytochrome C3 at Mercury Electrodes. *J. Electroanal. Chem. Interfacial Electrochem.* **1991**, *297*, 271–282. [CrossRef]

220. Ron, I.; Sepunaru, L.; Itzhakov, S.; Belenkova, T.; Friedman, N.; Pecht, I.; Sheves, M.; Cahen, D. Proteins as Electronic Materials: Electron Transport through Solid-State Protein Monolayer Junctions. *J. Am. Chem. Soc.* **2010**, *132*, 4131–4140. [CrossRef] [PubMed]

221. Sepunaru, L.; Pecht, I.; Sheves, M.; Cahen, D. Solid-State Electron Transport across Azurin: From a Temperature-Independent to a Temperature-Activated Mechanism. *J. Am. Chem. Soc.* **2011**, *133*, 2421–2423. [CrossRef] [PubMed]

222. Amdursky, N.; Marchak, D.; Sepunaru, L.; Pecht, I.; Sheves, M.; Cahen, D. Electronic Transport via Proteins. *Adv. Mater.* **2014**, *26*, 7142–7161. [CrossRef]

223. Valianti, S.; Cuevas, J.C.; Skourtis, S.S. Charge-Transport Mechanisms in Azurin-Based Monolayer Junctions. *J. Phys. Chem. C* **2019**, *123*, 5907–5922. [CrossRef]

224. Beratan, D.N.; Skourtis, S.S.; Balabin, I.A.; Balaeff, A.; Keinan, S.; Venkatramani, R.; Xiao, D. Steering Electrons on Moving Pathways. *Acc. Chem. Res.* **2009**, *42*, 1669–1678. [CrossRef]

225. Skourtis, S.S.; Waldeck, D.H.; Beratan, D.N. Fluctuations in Biological and Bioinspired Electron-Transfer Reactions. *Annu. Rev. Phys. Chem.* **2010**, *61*, 461–485. [CrossRef]

226. May, V.; Kühn, O. *Charge and Energy Transfer Dynamics in Molecular Systems*; Wiley-VCH Verlag GmbH & Co. KgaA: Weinheim, Germany, 2011.

227. Winkler, J.R.; Gray, H.B. Electron Flow through Metalloproteins. *Chem. Rev.* **2014**, *114*, 3369–3380. [CrossRef]

228. Beratan, D.N.; Liu, C.; Migliore, A.; Polizzi, N.F.; Skourtis, S.S.; Zhang, P.; Zhang, Y. Charge Transfer in Dynamical Biosystems, or The Treachery of (Static) Images. *Acc. Chem. Res.* **2015**, *48*, 474–481. [CrossRef] [PubMed]

229. Adams, E.M.; Lampret, O.; König, B.; Happe, T.; Havenith, M. Solvent Dynamics Play a Decisive Role in the Complex Formation of Biologically Relevant Redox Proteins. *Phys. Chem. Chem. Phys.* **2020**, *22*, 7451–7459. [CrossRef] [PubMed]

230. Venkat, A.S.; Corni, S.; Di Felice, R. Electronic Coupling Between Azurin and Gold at Different Protein/Substrate Orientations. *Small* **2007**, *3*, 1431–1437. [CrossRef] [PubMed]

231. Castañeda Ocampo, O.E.; Gordiichuk, P.; Catarci, S.; Gautier, D.A.; Herrmann, A.; Chiechi, R.C. Mechanism of Orientation-Dependent Asymmetric Charge Transport in Tunneling Junctions Comprising Photosystem I. *J. Am. Chem. Soc.* **2015**, *137*, 8419–8427. [CrossRef] [PubMed]

232. Amdursky, N.; Ferber, D.; Bortolotti, C.A.; Dolgikh, D.A.; Chertkova, R.V.; Pecht, I.; Sheves, M.; Cahen, D. Solid-State Electron Transport via Cytochrome c Depends on Electronic Coupling to Electrodes and across the Protein. *Proc. Natl. Acad. Sci. USA* **2014**, *111*, 5556–5561. [CrossRef]

233. Seeman, N.C. DNA in a Material World. *Nature* **2003**, *421*, 427–431. [CrossRef]

234. LaBean, T.H.; Li, H. Constructing Novel Materials with DNA. *Nano Today* **2007**, *2*, 26–35. [CrossRef]

235. Linko, V.; Paasonen, S.T.; Kuzyk, A.; Törmä, P.; Toppari, J.J. Characterization of the Conductance Mechanisms of DNA Origami by AC Ompedance Spectroscopy. *Small* **2009**, *5*, 2382–2386. [CrossRef] [PubMed]

236. Tuukkanen, S.; Toppari, J.J.; Hytönen, V.P.; Kuzyk, A.; Kulomaa, M.S.; Törmä, P. Dielectrophoresis as a Tool for Nanoscale DNA Manipulation. *Int. J. Nanotechnol.* **2005**, *2*, 280–291. [CrossRef]

237. Liu, Y.; Chung, J.H.; Liu, W.K.; Ruoff, R.S. Dielectrophoretic Assembly of Nanowires. *J. Phys. Chem. B* **2006**, *110*, 14098–14106. [CrossRef] [PubMed]

238. Krupke, R.; Hennrich, F.; Kappes, M.M.; Löhneysen, H.V. Surface Conductance Induced Dielectrophoresis of Semiconducting Single-Walled Carbon Nanotubes. *Nano Lett.* **2004**, *4*, 1395–1399. [CrossRef]

239. Vijayaraghavan, A.; Sciascia, C.; Dehm, S.; Lombardo, A.; Bonetti, A.; Ferrari, A.C.; Krupke, R. Dielectrophoretic Assembly of High-Density Arrays of Individual Graphene Devices for Rapid Screening. *ACS Nano* **2009**, *3*, 1729–1734. [CrossRef] [PubMed]

240. Freer, E.M.; Grachev, O.; Duan, X.; Martin, S.; Stumbo, D.P. High-Yield Self-Limiting Single-Nanowire Assembly with Dielectrophoresis. *Nat. Nanotechnol.* **2010**, *5*, 525–530. [CrossRef] [PubMed]

241. Schukfeh, M.I.; Sepunaru, L.; Behr, P.; Li, W.; Pecht, I.; Sheves, M.; Cahen, D.; Tornow, M. Towards Nanometer-Spaced Silicon Contacts to Proteins. *Nanotechnology* **2016**, *27*, 115302. [CrossRef] [PubMed]

242. Barik, A.; Zhang, Y.; Grassi, R.; Nadappuram, B.P.; Edel, J.B.; Low, T.; Koester, S.J.; Oh, S.-H.H. Graphene-Edge Dielectrophoretic Tweezers for Trapping of Biomolecules. *Nature* **2017**, *8*, 1867. [CrossRef]

243. He, X.; Tang, J.; Hu, H.; Shi, J.; Guan, Z.; Zhang, S.; Xu, H. Electrically Driven Optical Antennas Based on Template Dielectrophoretic Trapping. *ACS Nano* **2019**, *13*, 14041–14047. [CrossRef]

244. Hughes, M.P. AC Electrokinetics: Applications for Nanotechnology. *Nanotechnology* **2000**, *11*, 124–132. [CrossRef]

245. Pethig, R. Review Article-Dielectrophoresis: Status of the Theory, Technology, and Applications. *Biomicrofluidics* **2010**, *4*, 22811. [CrossRef]

246. Jones, T.B. Basic Theory of Dielectrophoresis and Electrorotation. *IEEE Eng. Med. Biol. Mag.* **2003**, *22*, 33–42. [CrossRef]

247. Castellanos, A.; Ramos, A.; González, A.; Green, N.G.; Morgan, H. Electrohydrodynamics and Dielectrophoresis in Microsystems: Scaling Laws. *J. Phys. D Appl. Phys.* **2003**, *36*, 2584–2597. [CrossRef]

248. Burke, P.J. Nanodielectrophoresis: Electronic Nanotweezers. In *Encyclopedia of Nanoscience and Nanotechnology*; Nalwa, H.S., Ed.; American Scientific Publishers: Valencia, CA, USA, 2003; pp. 1–19.

249. Lapizco-Encinas, B.H.; Rito-Palomares, M. Dielectrophoresis for the Manipulation of Nanobioparticles. *Electrophoresis* **2007**, *28*, 4521–4538. [CrossRef]

250. Zhang, C.; Khoshmanesh, K.; Mitchell, A.; Kalantar-zadeh, K. Dielectrophoresis for Manipulation of Micro/Nano Particles in Microfluidic Systems. *Anal. Bioanal. Chem.* **2010**, *396*, 401–420. [CrossRef] [PubMed]

251. Meighan, M.M.; Staton, S.J.R.; Hayes, M.A. Bioanalytical Separations Using Electric Field Gradient Techniques. *Electrophoresis* **2009**, *30*, 852–865. [CrossRef] [PubMed]

252. Kuzyk, A. Dielectrophoresis at the Nanoscale. *Electrophoresis* **2011**, *32*, 2307–2313. [CrossRef] [PubMed]

253. Ismael, A.; Wang, X.; Bennett, T.L.R.; Wilkinson, L.A.; Robinson, B.J.; Long, N.J.; Cohen, F.; Lambert, C.J. Tuning the Thermoelectrical Properties of Anthracene-Based Self-Assembled Monolayers. *Chem. Sci.* **2020**, *11*, 6836–6841. [CrossRef]

254. Wang, X.; Bennett, T.L.R.; Ismael, A.; Wilkinson, L.A.; Hamill, J.; White, A.J.P.; Grace, I.M.; Kolosov, O.V.; Albrecht, T.; Robinson, B.J.; et al. Scale-up of Room-Temperature Constructive Quantum Interference from Single Molecules to Self-Assembled Molecular-Electronic Films. *J. Am. Chem. Soc.* **2020**, *142*, 8555–8560. [CrossRef] [PubMed]

255. Xing, L.; Peng, Z.; Li, W.; Wu, K. On Controllability and Applicability of Surface Molecular Self-Assemblies. *Acc. Chem. Res.* **2019**, *52*, 1048–1058. [CrossRef]

256. Nickmans, K.; Schenning, A.P.H.J. Directed Self-Assembly of Liquid-Crystalline Molecular Building Blocks for Sub-5 Nm Nanopatterning. *Adv. Mater.* **2018**, *30*, 1703713. [CrossRef]

257. Bolu, B.S.; Sanyal, R.; Sanyal, A. Drug Delivery Systems from Self-Assembly of Dendron-Polymer Conjugates. *Molecules* **2018**, *23*, 1570. [CrossRef]

258. Song, Z.; Chen, X.; You, X.; Huang, K.; Dhinakar, A.; Gu, Z.; Wu, J. Self-Assembly of Peptide Amphiphiles for Drug Delivery: The Role of Peptide Primary and Secondary Structures. *Biomater. Sci.* **2017**, *5*, 2369–2380. [CrossRef]

259. Galeotti, F.; Pisco, M.; Cusano, A. Self-Assembly on Optical Fibers: A Powerful Nanofabrication Tool for next Generation "Lab-on-Fiber" Optrodes. *Nanoscale* **2018**, *10*, 22673–22700. [CrossRef] [PubMed]

260. Zhang, X.; Parekh, G.; Guo, B.; Huang, X.; Dong, Y.; Han, W.; Chen, X.; Xiao, G. Polyphenol and Self-Assembly: Metal Polyphenol Nanonetwork for Drug Delivery and Pharmaceutical Applications. *Future Drug Discov.* **2019**, *1*, FDD7. [CrossRef]

261. Hussain, S.A.; Dey, B.; Bhattacharjee, D.; Mehta, N. Unique Supramolecular Assembly through Langmuir—Blodgett (LB) Technique. *Heliyon* **2018**, *4*, e01038. [CrossRef] [PubMed]

262. Moehwald, H.; Brezesinski, G. From Langmuir Monolayers to Multilayer Films. *Langmuir* **2016**, *32*, 10445–10458. [CrossRef]

263. Dopierała, K.; Rojewska, M.; Skrzypiec, M.; Dutkiewicz, M.; Maciejewski, H.; Prochaska, K. Preparation and Characterisation of Monolayers and Langmuir–Blodgett Films of Liquid Crystal Mixed with Cubic Silsesquioxanes. *Liq. Cryst.* **2017**, *45*, 351–361. [CrossRef]

264. Pradilla, D.; Simon, S.; Sjöblom, J.; Samaniuk, J.; Skrzypiec, M.; Vermant, J. Sorption and Interfacial Rheology Study of Model Asphaltene Compounds. *Langmuir* **2016**, *32*, 2900–2911. [CrossRef]

265. Ma, Y.; Xie, Y.; Lin, L.; Zhang, L.; Liu, M.; Guo, Y.; Lan, Z.; Lu, Z. Photodimerization Kinetics of a Styrylquinoline Derivative in Langmuir–Blodgett Monolayers Monitored by Second Harmonic Generation. *J. Phys. Chem. C* **2017**, *121*, 23541–23550. [CrossRef]

266. Wang, Z.; Si, J.; Song, Z.; Zhang, P.; Wang, J.; Hao, Y.; Li, W.; Zhang, P.; Miao, S. Precise and Instrumental Measurement of Thermodynamics and Kinetics of Froth Flotation by Langmuir-Blodgett Technique. *Colloids Surf. A Physicochem. Eng. Asp.* **2020**, *605*, 125337. [CrossRef]

267. Da, C.; Junior, R.; Caseli, L. Adsorption and Enzyme Activity of Asparaginase at Lipid Langmuir and Langmuir-Blodgett Films. *Mater. Sci. Eng. C* **2017**, *73*, 579–584.

268. Brand, I.; Juhaniewicz-Debinska, J.; Wickramasinghe, L.; Verani, C.N. An in Situ Spectroelectrochemical Study on the Orientation Changes of an [FeIIILN2O3] Metallosurfactant Deposited as LB Films on Gold Electrode Surfaces. *Dalton Trans.* **2018**, *47*, 14218–14226. [CrossRef]

269. Su, Z.; Shodiev, M.; Leitch, J.J.; Abbasi, F.; Lipkowski, J. In Situ Electrochemical and PM-IRRAS Studies of Alamethicin Ion Channel Formation in Model Phospholipid Bilayers. *J. Electroanal. Chem.* **2018**, *819*, 251–259. [CrossRef]

270. Cea, P.; Martín, S.; Villares, A.; Möbius, D.; Carmen López, M. Use of UV-Vis Reflection Spectroscopy for Determining the Organization of Viologen and Viologen Tetracyanoquinodimethanide Monolayers. *J. Phys. Chem. B* **2006**, *110*, 963–970. [CrossRef] [PubMed]

271. Ariga, K.; Yamauchi, Y.; Mori, T.; Hill, J.P. 25th Anniversary Article: What Can Be Done with the Langmuir-Blodgett Method? Recent Developments and Its Critical Role in Materials Science. *Adv. Mater.* **2013**, *25*, 6477–6512. [CrossRef] [PubMed]

272. Hussain, S.A.; Bhattacharjee, D. Langmuir Blodgett Films and Molecular Electronics. *Mod. Phys. Lett. B* **2009**, *23*, 3437–3451. [CrossRef]

273. Hirahara, M.; Kaida, H.; Miyauchi, Y.; Goto, H.; Yamagishi, A.; Umemura, Y. Application of Electrospray Spreading to a Modified Langmuir-Blodgett Technique for Organo-Clay Hybrid Film Preparation. *Colloids Surf. A* **2019**, *580*, 123714. [CrossRef]

274. Giner-casares, J.J.; Brezesinski, G.; Möhwald, H. Current Opinion in Colloid & Interface Science Langmuir Monolayers as Unique Physical Models. *Curr. Opin. Colloid Interface Sci.* **2014**, *19*, 176–182.

275. Gyepi-Garbrah, S.H.; Šilerová, R. The First Direct Comparison of Self-Assembly and Langmuir-Blodgett Deposition Techniques: Two Routes to Highly Organized Monolayers. *Phys. Chem. Chem. Phys.* **2002**, *4*, 3436–3442. [CrossRef]

276. Villares, A.; Lydon, D.P.; Low, P.J.; Robinson, B.J.; Ashwell, G.J.; Royo, F.M.; Cea, P. Characterization and Conductivity of Langmuir–Blodgett Films Prepared from an Amine-Substituted Oligo (Phenylene Ethynylene). *Chem Mater.* **2008**, *20*, 258–264. [CrossRef]

277. Martín, S.; Pera, G.; Ballesteros, L.M.; Hope, A.J.; Marqués-González, S.; Low, P.J.; Pérez-Murano, F.; Nichols, R.J.; Cea, P. Towards the Fabrication of the Top-Contact Electrode in Molecular Junctions by Photoreduction of a Metal Precursor. *Chem. Eur. J.* **2014**, *20*, 3421–3426. [CrossRef]

278. Haro, M.; Giner, B.; Gascón, I.; Royo, F.M.; López, M.C. Isomerization Behavior of an Azopolymer in Terms of the Langmuir-Blodgett Film Thickness and the Transference Surface Pressure. *Macromolecules* **2007**, *40*, 2058–2069. [CrossRef]

279. Haro, M.; Gascón, I.; Aroca, R.; López, M.C.; Royo, F.M. Structural Characterization and Properties of an Azopolymer Arranged in Langmuir and Langmuir-Blodgett Films. *J. Colloid Interface Sci.* **2008**, *319*, 277–286. [CrossRef] [PubMed]

280. Jayamurugan, G.; Gowri, V.; Hernández, D.; Martin, S.; González-Orive, A.; Dengiz, C.; Dumele, O.; Pérez-Murano, F.; Gisselbrecht, J.P.; Boudon, C.; et al. Design and Synthesis of Aviram–Ratner-Type Dyads and Rectification Studies in Langmuir–Blodgett (LB) Films. *Chem. Eur. J.* **2016**, *22*, 10539–10547. [CrossRef]

281. Kausar, A. Survey on Langmuir-Blodgett Films of Polymer and Polymeric Composite. *Polym. Plast. Technol. Eng.* **2018**, *56*, 932–945. [CrossRef]

282. Silva, E.A.; Gregori, A.; Fernandes, J.D.; Njel, C.; DedryvèreDedryvère, R.; L Constantino, C.J.; Hiorns, R.C.; Lartigau-Dagron, C.; Olivati, C.A. Understanding the Langmuir and Langmuir-Schaefer Film Conformation of Low-Bandgap Polymers and Their Bulk Heterojunctions with PCBM. *Nanotechnology* **2020**, *31*, 315712–315726. [CrossRef] [PubMed]

283. Yamamoto, S.; Nishina, N.; Matsui, J.; Miyashita, T.; Mitsuishi, M. High-Density and Monolayer-Level Integration of π-Conjugated Units: Amphiphilic Carbazole Homopolymer Langmuir-Blodgett Films. *Langmuir* **2018**, *34*, 10491–10497. [CrossRef] [PubMed]

284. Dey, B.; Chakraborty, S.; Chakraborty, S.; Bhattacharjee, D.; Khan, A.; Arshad Hussain, S. Electrical Switching Behaviour of a Metalloporphyrin in Langmuir-Blodgett Film. *Org. Electron.* **2018**, *55*, 50–62. [CrossRef]

285. Nizioł, J.; Makyła-Juzak, K.; Radko, A.; Ekiert, R.; Zemła, J.; Górska, N.; Chachaj-Brekiesz, A.; Marzec, M.; Harańczyk, H.; Dynarowicz-Latka, P. Linear, Self-Assembled Patterns Appearing Spontaneously as a Result of DNA-CTMA Lipoplex Langmuir-Blodgett Deposition on a Solid Surface. *Polymer* **2019**, *178*, 121643. [CrossRef]

286. Makowiecki, J.; Piosik, E.; Neunert, G.; Stolarski, R.; Piecek, W.; Martynski, T. Molecular Organization of Perylene Derivatives in Langmuir-Blodgett Multilayers. *Opt. Mater.* **2015**, *46*, 555–560. [CrossRef]

287. Diego Fernandes, J.; Maximino, M.D.; Braunger, M.L.; Pereira, M.S.; de Almeida Olivati, C.; Constantino, C.J.L.; Alessio, P. Supramolecular Architecture and Electrical Conductivity in Organic Semiconducting Thin Films. *Phys. Chem. Chem. Phys.* **2020**, *22*, 13554–13562. [CrossRef]

288. Cavazzini Cesca, E.; Maria Hoffmeister, D.; Naidek, K.P.; Batista Marques Novo, A.; Serbena, J.P.; Hümmelgen, I.A.; Westphal, E.; Araki, K.; Toma, H.E.; Winnischofer, H. 1,3,4-Oxadiazole Based Ruthenium Amphiphile for Langmuir-Blodgett Films and Photo-Responsive Logic Gate Construction. *Electrochim. Acta* **2020**, *350*, 136350. [CrossRef]

289. Ohashi, T.; Kikuchi, N.; Fujimori, A. Creation of Highly Ordered "Nano-Mille-Feuille" Hard/Soft Nanoparticle Multilayers with Interparticle Cross-Linking by Diacetylene-Containing Chains. *Langmuir* **2020**, *36*, 5596–5607. [CrossRef] [PubMed]

290. Kursunlu, A.N.; Acikbas, Y.; Ozmen, M.; Erdogan, M.; Capan, R. Fabrication of LB Thin Film of Pillar[5]Arene-2-Amino-3-Hydroxypyridine for the Sensing of Vapors. *Mater. Lett.* **2020**, *267*, 127538. [CrossRef]

291. Liu, H.; Siron, M.; Gao, M.; Lu, D.; Bekenstein, Y.; Zhang, D.; Dou, L.; Alivisatos, A.P.; Yang, P. Lead Halide Perovskite Nanowires Stabilized by Block Copolymers for Langmuir-Blodgett Assembly. *Nano Res.* **2020**, *13*, 1453–1458. [CrossRef]

292. Su, L.; Xu, F.; Chen, J.; Cao, Y.; Wang, C. Photoresponsive 2D Polymeric Langmuir–Blodgett Films of 2,3,6,7,10,11-Hexaiminotriphenylene. *New J. Chem.* **2020**, *44*, 5656–5660. [CrossRef]

293. Ariga, K.; Ji, Q.; Nakanishi, W.; Hill, J.P.; Aono, M. Nanoarchitectonics: A New Materials Horizon for Nanotechnology. *Mater. Horiz.* **2015**, *2*, 406–413. [CrossRef]

294. Ariga, K.; Yamauchi, Y. Nanoarchitectonics from Atom to Life. *Chem. Asian J.* **2020**, *15*, 718–728. [CrossRef]

295. Ariga, K. Don't Forget Langmuir-Blodgett Films 2020: Interfacial Nanoarchitectonics with Molecules, Materials, and Living Objects. *Langmuir* **2020**, *36*, 7158–7180. [CrossRef]

296. Ariga, K.; Mori, T.; Li, J. Langmuir Nanoarchitectonics from Basic to Frontier. *Langmuir* **2019**, *35*, 3585–3599. [CrossRef]

297. Govindaraju, T.; Avinash, M.B. Two-Dimensional Nanoarchitectonics: Organic and Hybrid Materials. *Nanoscale* **2012**, *4*, 6102–6117. [CrossRef]

298. Ramanathan, M.; Shrestha, L.K.; Mori, T.; Ji, Q.; Hill, J.P.; Ariga, K. Amphiphile Nanoarchitectonics: From Basic Physical Chemistry to Advanced Applications. *Phys. Chem. Chem. Phys.* **2013**, *15*, 10580–10611. [CrossRef]

299. Ariga, K.; Mori, T.; Kitao, T.; Uemura, T. Supramolecular Chiral Nanoarchitectonics. *Adv. Mater.* **2020**, *2020*, 1905657. [CrossRef] [PubMed]

300. Komiyama, M.; Mori, T.; Ariga, K. Molecular Imprinting: Materials Nanoarchitectonics with Molecular Information. *Bull. Chem. Soc. Jpn.* **2018**, *91*, 1075–1111. [CrossRef]

301. Liu, J.; Zhou, H.; Yang, W.; Ariga, K. Soft Nanoarchitectonics for Enantioselective Biosensing. *Acc. Chem. Res.* **2020**, *53*, 644–653. [CrossRef] [PubMed]

302. Abe, H.; Liu, J.; Ariga, K. Catalytic Nanoarchitectonics for Environmentally Compatible Energy Generation. *Mater. Today* **2016**, *19*, 12–18. [CrossRef]

303. Komiyama, M.; Ariga, K. Nanoarchitectonics to Prepare Practically Useful Artificial Enzymes. *Mol. Catal.* **2019**, *475*, 110492. [CrossRef]

304. Ariga, K.; Ji, Q.; Mori, T.; Naito, M.; Yamauchi, Y.; Abe, H.; Hill, J.P. Enzyme Nanoarchitectonics: Organization and Device Application. *Chem. Soc. Rev.* **2013**, *42*, 6322–6345. [CrossRef]

305. Ariga, K.; Ito, M.; Mori, T.; Watanabe, S.; Takeya, J. Atom/Molecular Nanoarchitectonics for Devices and Related Applications. *Nano Today* **2019**, *28*, 100762. [CrossRef]

306. Xu, J.; Zhang, J.; Zhang, W.; Lee, C.S. Interlayer Nanoarchitectonics of Two-Dimensional Transition-Metal Dichalcogenides Nanosheets for Energy Storage and Conversion Applications. *Adv. Energy Mater.* **2017**, *7*, 1700571. [CrossRef]

307. Nakanishi, W.; Minami, K.; Shrestha, L.K.; Ji, Q.; Hill, J.P.; Ariga, K. Bioactive Nanocarbon Assemblies: Nanoarchitectonics and Applications. *Nano Today* **2014**, *9*, 378–394. [CrossRef]

308. Ariga, K.; Leong, D.T.; Mori, T. Nanoarchitectonics for Hybrid and Related Materials for Bio-Oriented Applications. *Adv. Funct. Mater.* **2018**, *28*, 1702905. [CrossRef]

309. Liang, X.; Li, L.; Tang, J.; Komiyama, M.; Ariga, K. Dynamism of Supramolecular DNA/RNA Nanoarchitectonics: From Interlocked Structures to Molecular Machines. *Bull. Chem. Soc. Jpn.* **2020**, *93*, 581–603. [CrossRef]

310. Chiodini, S.; Ruiz-Rincón, S.; Garcia, P.D.; Martin, S.; Kettelhoit, K.; Armenia, I.; Werz, D.B.; Cea, P. Bottom Effect in Atomic Force Microscopy Nanomechanics. *Small* **2020**, *2020*, 2000269. [CrossRef]

311. Rauf, S.; Vijjapu, M.T.; Andrés, M.A.; Gascón, I.; Roubeau, O.; Eddaoudi, M.; Salama, K.N. Highly Selective Metal–Organic Framework Textile Humidity Sensor. *ACS Appl. Mater. Interfaces* **2020**, *12*, 29999–30006. [CrossRef] [PubMed]

312. Bélanger, D.; Pinson, J. Electrografting: A Powerful Method for Surface Modification. *Chem. Soc. Rev.* **2011**, *40*, 3995–4048. [CrossRef] [PubMed]

313. Li, W.; Liu, Q.; Zhang, Y.; Li, C.; He, Z.; Choy, W.C.H.; Low, P.J.; Sonar, P.; Kyaw, A.K.K. Biodegradable Materials and Green Processing for Green Electronics. *Adv. Mater.* **2020**, *2020*, 2001591. [CrossRef]

314. Hapiot, P.; Lagrost, C.; Leroux, Y.R. Molecular Nano-Structuration of Carbon Surfaces through Reductive Diazonium Salts Grafting. *Curr. Opin. Electrochem.* **2018**, *7*, 103–108. [CrossRef]

315. Pinson, J.; Podvorica, F.I. Surface Modification of Materials: Electrografting of Organic Films. *Curr. Opin. Electrochem.* **2020**, *24*, 44–48. [CrossRef]

316. Delamar, M.; Hitmi, R.; Pinson, J.; Savéant, J. Covalent Modification of Carbon Surfaces by Grafting of Functionalized Aryl Radicals Produced from Electrochemical Reduction of Diazonium Salts. *J. Am. Chem. Soc.* **1992**, *114*, 5883–5884. [CrossRef]

317. Anex, C.; Touzé, E.; Curet, L.; Gohier, F.; Cougnon, C. Base-Assisted Electrografting of Aromatic Amines. *ChemElectroChem* **2019**, *6*, 4963–4969. [CrossRef]

318. Mccreery, R.L. The Merger of Electrochemistry and Molecular Electronics. *Chem. Rec.* **2012**, *12*, 149–163. [CrossRef]

319. Christophe Lacroix, J. Electrochemistry Does the Impossible: Robust and Reliable Large Area Molecular Junctions. *Curr. Opin. Electrochem.* **2018**, *7*, 153–160. [CrossRef]

320. Randriamahazaka, H.; Ghilane, J. Electrografting and Controlled Surface Functionalization of Carbon Based Surfaces for Electroanalysis. *Electroanalysis* **2016**, *28*, 13–26. [CrossRef]

321. Yan, H.; Bergren, A.J.; McCreery, R.L. All-Carbon Molecular Tunnel Junctions. *J. Am. Chem. Soc.* **2011**, *133*, 19168–19177. [CrossRef] [PubMed]

322. McCreery, R.; Bergren, A.; Morteza-Najarian, A.; Sayed, S.Y.; Yan, H. Electron Transport in All-Carbon Molecular Electronic Devices. *Faraday Discuss.* **2014**, *172*, 9–25. [CrossRef] [PubMed]

323. Sayed, S.Y.; Bayat, A.; Kondratenko, M.; Leroux, Y.; Hapiot, P.; McCreery, R.L. Bilayer Molecular Electronics: All-Carbon Electronic Junctions Containing Molecular Bilayers Made with "Click" Chemistry. *J. Am. Chem. Soc.* **2013**, *135*, 12972–12975. [CrossRef] [PubMed]

324. Guselnikova, O.; Miliutina, E.; Elashnikov, R.; Burtsev, V.; Chehimi, M.M.; Svorcik, V.; Yusubov, M.; Lyutakov, O.; Postnikov, P. Chemical Modification of Gold Surface via UV-Generated Aryl Radicals Derived 3,5-Bis(Trifluoromethyl)Phenyl)Iodonium Salt. *Prog. Org. Coat.* **2019**, *136*, 105211. [CrossRef]

325. Ramírez-Chan, D.E.; Fragoso-Soriano, R.; González, F.J. Effect of Electrolyte Ions on the Formation, Electroactivity, and Rectification Properties of Films Obtained by Electrografting. *ChemElectroChem* **2020**, *7*, 904–913. [CrossRef]

326. Madsen, M.R.; Koefoed, L.; Jensen, H.; Daasbjerg, K.; Pedersen, S.U. Two-Phase Bipolar Electrografting. *Electrochim. Acta* **2019**, *317*, 61–69. [CrossRef]

327. Arrotin, B.; Noël, J.M.; Delhalle, J.; Mespouille, L.; Mekhalif, Z. Electrografting of Mixed Organophosphonic Monolayers for SI-ATRP of 2-Methacryloyloxyethyl Phosphorylcholine. *J. Coat. Technol. Res.* **2019**, *16*, 1121–1132. [CrossRef]

328. Guo, D.; Wang, L.; Wang, X.; Xiao, Y.; Wang, C.; Chen, L.; Ding, Y. PEDOT Coating Enhanced Electromechanical Performances and Prolonged Stable Working Time of IPMC Actuator. *Sens. Actuators B Chem.* **2020**, *305*, 127488. [CrossRef]

329. Villemin, E.; Lemarque, B.; Thiêt Vû, T.; Nguyen, V.Q.; Trippé-Allard, G.; Martin, P.; Lacaze, P.-C.; Lacroix, J.-C. Improved Adhesion of Poly(3,4-Ethylenedioxythiophene) (PEDOT) Thin Film to Solid Substrates Using Electrografted Promoters and Application to Efficient Nanoplasmonic Devices. *Synth. Met.* **2019**, *248*, 45–52. [CrossRef]

330. Gutiérrez-Sánchez, C.; Mediavilla, O.; Guerrero-Esteban, T.; Revenga-Parra, O.; Pariente, E.; On Lorenzo, E. Direct Covalent Immobilization of New Nitrogen-Doped Carbon Nanodots by Electrografting for Sensing Applications. *Carbon* **2020**, *159*, 303–310. [CrossRef]

331. Berg, K.E.; Leroux, Y.R.; Hapiot, P.; Henry, C.S. Increasing Applications of Graphite Thermoplastic Electrodes with Aryl Diazonium Grafting. *ChemElectroChem* **2019**, *6*, 4811–4816. [CrossRef]

332. Aceta, Y.; Leroux, Y.R.; Hapiot, P. Evaluation of Alkyl-Ferrocene Monolayers on Carbons for Charge Storage Applications, a Voltammetry and Impedance Spectroscopy Investigation. *ChemElectroChem* **2019**, *6*, 1704–1710. [CrossRef]

333. Wu, T.; Lankshear, E.R.; Downard, A.J. Simultaneous Electro-Click and Electrochemically Mediated Polymerization Reactions for One-Pot Grafting from a Controlled Density of Anchor Sites. *ChemElectroChem* **2019**, *6*, 5149–5154. [CrossRef]

334. Gabaji, M.; Médard, J.M.; Hemmerle, A.; Pinson, J.; Michel, J.P. From Langmuir–Blodgett to Grafted Films. *Langmuir* **2020**, *36*, 2534–2542. [CrossRef]

335. Dalla Francesca, K.; Salhani, C.; Timpa, S.; Rastikian, J.; Suffit, S.; Martin, P.; Lacroix, J.; Lafarge, P.; Barraud, C.; Della Rocca, M.L. Large-Area in Plane Molecular Junctions by Electrografting in 10 Nm Metallic Nanotrenches. *AIP Adv.* **2020**, *10*, 25023. [CrossRef]

336. Nguyen, H.H.; Park, J.; Kang, S.; Kim, M. Surface Plasmon Resonance: A Versatile Technique for Biosensor Applications. *Sensors* **2015**, *15*, 10481–10510. [CrossRef]

337. Fioresi, F.; Rouleau, A.; Maximova, K.; Vieillard, J.; Boireau, W.; Caille, C.E.; Soulignac, C.; Zeggari, R.; Clamens, T.; Lesouhaitier, O.; et al. Electrografting of Diazonium Salt for SPR Application. *Mater. Today Proc.* **2019**, *6*, 340–344. [CrossRef]

338. Koefoed, L.; Pedersen, S.U.; Daasbjerg, K. Bipolar Electrochemistry—A Wireless Approach for Electrode Reactions. *Curr. Opin. Electrochem.* **2017**, *2*, 13–17. [CrossRef]

339. Available online: http://www.cabot-corp.com (accessed on 14 August 2020).

340. Levesque, L.; Lawrence, M.F.; Bourguignon, B.; Leclerc, G. Drug Eluting Device for Treating Vascular Diseases. WO Patent WO 2002066092, 19 August 2020.

341. Nichols, R.J.; Haiss, W.; Higgins, S.J.; Leary, E.; Martin, S.; Bethell, D. The Experimental Determination of the Conductance of Single Molecules. *Phys. Chem. Chem. Phys.* **2010**, *12*, 2801–2815. [CrossRef] [PubMed]

342. Walker, A.V. Building Robust and Reliable Molecular Constructs: Patterning, Metallic Contacts, and Layer-by-Layer Assembly. *Langmuir* **2010**, *26*, 13778–13785. [CrossRef] [PubMed]

343. Wohlfart, P.; Weiß, J.; Käshammer, J.; Kreiter, M.; Winter, C.; Fischer, R.; Mittler-Neher, S. MOCVD of Aluminum Oxide/Hydroxide onto Organic Self-Assembled Monolayers. *Chem. Vap. Depos.* **1999**, *5*, 165–170. [CrossRef]

344. Haick, H.; Cahen, D. Contacting Organic Molecules by Soft Methods: Towards Molecule-Based Electronic Devices. *Acc. Chem. Res.* **2008**, *41*, 359–366. [CrossRef] [PubMed]

345. Hansen, C.R.; Sørensen, T.J.; Glyvradal, M.; Larsen, J.; Eisenhardt, S.H.; Bjørnholm, T.; Nielsen, M.M.; Feidenhans'l, R.; Laursen, B.W. Structure of the Buried Metal-Molecule Interface in Organic Thin Film Devices. *Nano Lett.* **2009**, *9*, 1052–1057. [CrossRef] [PubMed]

346. Haick, H.; Niitsoo, O.; Ghabboun, J.; Cahen, D. Electrical Contacts to Organic Molecular Films by Metal Evaporation: Effect of Contacting Details. *J. Phys. Chem. C* **2007**, *111*, 2318–2329. [CrossRef]

347. Haick, H.; Ambrico, M.; Ghabboun, J.; Ligonzo, T.; Cahen, D. Contacting Organic Molecules by Metal Evaporation. *Phys. Chem. Chem. Phys.* **2004**, *6*, 4538–4541. [CrossRef]

348. Xu, T.; Peterson, I.R.; Lakshmikantham, M.V.; Metzger, R.M. Rectification by a Monolayer of Hexadecylquinolinium Tricyanoquinodimethanide between Gold Electrodes. *Angew. Chem. Int. Ed.* **2001**, *40*, 1749–1752. [CrossRef]

349. Lodha, S.; Janes, D.B. Metal/Molecule/p-Type GaAs Heterostructure Devices. *J. Appl. Phys.* **2006**, *100*, 024503. [CrossRef]

350. Daniel, M.C.; Astruc, D. Gold Nanoparticles: Assembly, Supramolecular Chemistry, Quantum-Size-Related Properties, and Applications Toward Biology, Catalysis, and Nanotechnology. *Chem. Rev.* **2004**, *104*, 293–346. [CrossRef]

351. Sebastián, V.; Smith, C.D.; Jensen, K.F. Shape-Controlled Continuous Synthesis of Metal Nanostructures. *Nanoscale* **2016**, *8*, 7534–7543. [CrossRef] [PubMed]

352. Huang, X.; Tang, S.; Mu, X.; Dai, Y.; Chen, G.; Zhou, Z.; Ruan, F.; Yang, Z.; Zheng, N. Freestanding Palladium Nanosheets with Plasmonic and Catalytic Properties. *Nat. Nanotechnol.* **2011**, *6*, 28–32. [CrossRef] [PubMed]

353. Bonifas, A.P.; McCreery, R.L. 'Soft' Au, Pt and Cu Contacts for Molecular Junctions through Surface-Diffusion-Mediated Deposition. *Nat. Nanotechnol.* **2010**, *5*, 612–617. [CrossRef] [PubMed]

354. Bonifas, A.P.; McCreery, R.L. Assembling Molecular Electronic Junctions One Molecule at a Time. *Nano Lett.* **2011**, *11*, 4725–4729. [CrossRef]

355. George, S.M. Atomic Layer Deposition: An Overview. *Chem. Rev.* **2010**, *110*, 111–131. [CrossRef]

356. Wu, J.Z.; Acharya, J.; Goul, R. In Vacuo Atomic Layer Deposition and Electron Tunneling Characterization of Ultrathin Dielectric Films for Metal/Insulator/Metal Tunnel Junctions. *J. Vac. Sci. Technol. A* **2020**, *38*, 040802. [CrossRef]

357. Seitz, O.; Dai, M.; Wallace, R.M.; Chabal, Y.J. Copper-Metal Deposition on Self Assembled Monolayer for Making Top Contacts in Molecular Electronic Devices. *J. Am. Chem. Soc.* **2009**, *131*, 18159–18167. [CrossRef]

358. Akkerman, H.B.; Blom, P.W.M.; De Leeuw, D.M.; De Boer, B. Towards Molecular Electronics with Large-Area Molecular Junctions. *Nature* **2006**, *441*, 69–72. [CrossRef]

359. Silien, C.; Buck, M. On the Role of Extrinsic and Intrinsic Defects in the Underpotential Deposition of Cu on Thiol-Modified Au(111) Electrodes. *J. Phys. Chem. C* **2008**, *112*, 3881–3890. [CrossRef]

360. Qu, D.; Uosaki, K. Formation of Continuous Platinum Layer on Top of an Organic Monolayer by Electrochemical Deposition Followed by Electroless Deposition. *J. Electroanal. Chem.* **2011**, *662*, 80–86. [CrossRef]

361. Eberle, F.; Saitner, M.; Boyen, H.G.; Kucera, J.; Gross, A.; Romanyuk, A.; Oelhafen, P.; D'Olieslaeger, M.; Manolova, M.; Kolb, D.M. A Molecular Double Decker: Extending the Limits of Current Metal-Molecule Hybrid Structures. *Angew. Chem. Int. Ed.* **2010**, *49*, 341–345. [CrossRef] [PubMed]

362. Eberle, F.; Metzler, M.; Kolb, D.M.; Saitner, M.; Wagner, P.; Boyen, H.G. Metallization of Ultra-Thin, Non-Thiol SAMs with Flat-Lying Molecular Units: Pd on 1, 4-Dicyanobenzene. *ChemPhysChem* **2010**, *11*, 2951–2956. [CrossRef] [PubMed]

363. Popoff, R.T.W.; Zavareh, A.A.; Kavanagh, K.L.; Yu, H.Z.; Popo, R.T.W.; Zavareh, A.A.; Kavanagh, K.L.; Yu, H.Z. Reduction of Gold Penetration through Phenyl-Terminated Alkyl Monolayers on Silicon. *J. Phys. Chem. C* **2012**, *116*, 17040–17047. [CrossRef]

364. Manolova, M.; Ivanova, V.; Kolb, D.M.; Boyen, H.G.; Ziemann, P.; Büttner, M.; Romanyuk, A.; Oelhafen, P. Metal Deposition onto Thiol-Covered Gold: Platinum on a 4-Mercaptopyridine SAM. *Surf. Sci.* **2005**, *590*, 146–153. [CrossRef]

365. Manolova, M.; Kayser, M.; Kolb, D.M.; Boyen, H.G.; Ziemann, P.; Mayer, D.; Wirth, A. Rhodium Deposition onto a 4-Mercaptopyridine SAM on Au (111). *Electrochim. Acta* **2007**, *52*, 2740–2745. [CrossRef]

366. Schlesinger, M.; Paunovic, M. *In Electrochemical Society Series*; Wiley: New York, NY, USA, 2000.

367. Aldakov, D.; Bonnassieux, Y.; Geffroy, B.; Palacin, S. Selective Electroless Copper Deposition on Self-Assembled Dithiol Monolayers. *ACS Appl. Mater. Interfaces* **2009**, *1*, 584–589. [CrossRef]

368. Shi, Z.; Walker, A.V. Synthesis of Nickel Nanowires via Electroless Nanowire Deposition on Micropatterned Substrates. *Langmuir* **2011**, *27*, 11292–11295. [CrossRef]

369. Zangmeister, C.D.; Van Zee, R.D. Electroless Deposition of Copper onto 4-Mercaptobenzoic Acid Self-Assembled on Gold. *Langmuir* **2003**, *19*, 8065–8068. [CrossRef]

370. Lu, P.; Shi, Z.; Walker, A.V. Selective Electroless Deposition of Copper on Organic Thin Films with Improved Morphology. *Langmuir* **2011**, *27*, 13022–13028. [CrossRef]

371. Lu, P.; Walker, A.V. Investigation of the Mechanism of Electroless Deposition of Copper on Functionalized Alkanethiolate Self-Assembled Monolayers Adsorbed on Gold. *Langmuir* **2007**, *23*, 12577–12582. [CrossRef]

372. Loo, Y.L.; Lang, D.V.; Rogers, J.A.; Hsu, J.W.P. Electrical Contacts to Molecular Layers by Nanotransfer Printing. *Nano Lett.* **2003**, *3*, 913–917. [CrossRef]

373. Niskala, J.R.; Rice, W.C.; Bruce, R.C.; Merkel, T.J.; Tsui, F.; You, W. Tunneling Characteristics of Au-Alkanedithiol-Au Junctions Formed via Nanotransfer Printing (NTP). *J. Am. Chem. Soc.* **2012**, *134*, 12072–12082. [CrossRef] [PubMed]

374. Guerin, D.; Merckling, C.; Lenfant, S.; Wallart, X.; Pleutin, S.; Vuillaume, D. Silicon-Molecules-Metal Junctions by Transfer Printing: Chemical Synthesis and Electrical Properties. *J. Phys. Chem. C* **2007**, *111*, 7947–7956. [CrossRef]

375. Shimizu, K.T.; Fabbri, J.D.; Jelincic, J.J.; Melosh, N.A. Soft Deposition of Large-Area Metal Contacts for Molecular Electronics. *Adv. Mater.* **2006**, *18*, 1499–1504. [CrossRef]

376. Majumdar, N.; Gergel, N.; Routenberg, D.; Bean, J.C.; Harriott, L.R.; Li, B.; Pu, L.; Yao, Y.; Tour, J.M. Nanowell Device for the Electrical Characterization of Metal-Molecule-Metal Junctions. *J. Vac. Sci. Technol. B Microelectron. Nanom. Struct.* **2005**, *23*, 1417–1421. [CrossRef]

377. Chen, J.; Reed, M.A.; Rawlett, A.M.; Tour, J.M. Large On-off Ratios and Negative Differential Resistance in a Molecular Electronic Device. *Science* **1999**, *286*, 1550–1552. [CrossRef] [PubMed]

378. Karuppannan, S.K.; Hongting, H.; Troadec, C.; Vilan, A.; Nijhuis, C.A. Ultrasmooth and Photoresist-Free Micropore-Based EGaIn Molecular Junctions: Fabrication and How Roughness Determines Voltage Response. *Adv. Funct. Mater.* **2019**, *29*, 1904452. [CrossRef]

379. Wang, G.; Kim, Y.; Choe, M.; Kim, T.W.; Lee, T. A New Approach for Molecular Electronic Junctions with a Multilayer Graphene Electrode. *Adv. Mater.* **2011**, *23*, 755–760. [CrossRef]

380. He, J.; Chen, B.; Flatt, A.K.; Stephenson, J.J.; Doyle, C.D.; Tour, J.M. Metal-Free Silicon-Molecule-Nanotube Testbed and Memory Device. *Nat. Mater.* **2006**, *5*, 63–68. [CrossRef]

381. Puebla-Hellmann, G.; Venkatesan, K.; Mayor, M.; Lörtscher, E. Metallic Nanoparticle Contacts for High-Yield, Ambient-Stable Molecular-Monolayer Devices. *Nature* **2018**, *559*, 232–235. [CrossRef]

382. Preiner, M.J.; Melosh, N.A. Creating Large Area Molecular Electronic Junctions Using Atomic Layer Deposition. *Appl. Phys. Lett.* **2008**, *92*, 213301. [CrossRef]

383. Milani, F.; Grave, C.; Ferri, V.; Samorì, P.; Rampi, M.A. Ultrathin π-Conjugated Polymer Films for Simple Fabrication of Large-Area Molecular Junctions. *ChemPhysChem* **2007**, *8*, 515–518. [CrossRef] [PubMed]

384. Li, T.; Hauptmann, J.R.; Wei, Z.; Petersen, S.; Bovet, N.; Vosch, T.; Nygård, J.; Hu, W.; Liu, Y.; Bjørnholm, T.; et al. Solution-Processed Ultrathin Chemically Derived Graphene Films as Soft Top Contacts for Solid-State Molecular Electronic Junctions. *Adv. Mater.* **2012**, *24*, 1333–1339. [CrossRef]

385. Jeong, H.; Kim, D.; Xiang, D.; Lee, T. High-Yield Functional Molecular Electronic Devices. *ACS Nano* **2017**, *11*, 6511–6548. [CrossRef] [PubMed]

386. Karuppannan, S.K.; Neoh, E.H.L.; Vilan, A.; Nijhuis, C.A. Protective Layers Based on Carbon Paint to Yield High-Quality Large-Area Molecular Junctions with Low Contact Resistance. *J. Am. Chem. Soc.* **2020**, *142*, 3513–3524. [CrossRef] [PubMed]

387. Nijhuis, C.A.; Reus, W.F.; Barber, J.R.; Dickey, M.D.; Whitesides, G.M. Charge Transport and Rectification in Arrays of SAM-Based Tunneling Junctions. *Nano Lett.* **2010**, *10*, 3611–3619. [CrossRef]

388. Jeong, I.; Song, H. Fabrication and Characterization of Graphene/Molecule/Graphene Vertical Junctions with Aryl Alkane Monolayers. *J. Korean Phys. Soc.* **2017**, *71*, 692–696. [CrossRef]

389. Guo, X.; Small, J.P.; Klare, J.E.; Wang, Y.; Purewal, M.S.; Tam, I.W.; Hong, B.H.; Caldwell, R.; Huang, L.; O'Brien, S.; et al. Covalently Bridging-Gaps in Single-Walled Carbon Nanotubes with Conducting Molecules. *Science* **2006**, *311*, 356–359. [CrossRef]

390. Cao, Y.; Dong, S.; Liu, S.; He, L.; Gan, L.; Yu, X.; Steigerwald, M.L.; Wu, X.; Liu, Z.; Guo, X. Building High-Throughput Molecular Junctions Using Indented Graphene Point Contacts. *Angew. Chem.* **2012**, *124*, 12394–12398. [CrossRef]

391. Supur, M.; Van Dyck, C.; Bergren, A.J.; McCreery, R.L. Bottom-up, Robust Graphene Ribbon Electronics in All-Carbon Molecular Junctions. *ACS Appl. Mater. Interfaces* **2018**, *10*, 6090–6095. [CrossRef]

392. Tefashe, U.M.; Van Dyck, C.; Saxena, S.K.; Lacroix, J.C.; McCreery, R.L. Unipolar Injection and Bipolar Transport in Electroluminescent Ru-Centered Molecular Electronic Junctions. *J. Phys. Chem. C* **2019**, *123*, 29162–29172. [CrossRef]

393. Najarian, A.M.; McCreery, R.L. Long-Range Activationless Photostimulated Charge Transport in Symmetric Molecular Junctions. *ACS Nano* **2019**, *13*, 867–877. [CrossRef] [PubMed]

Review

Attenuation Factors in Molecular Electronics: Some Theoretical Concepts

Yannick J. Dappe

SPEC, CEA, CNRS, Université Paris-Saclay, CEA Saclay, 91191 Gif-sur-Yvette, CEDEX, France; yannick.dappe@cea.fr

Received: 30 July 2020; Accepted: 1 September 2020; Published: 4 September 2020

Abstract: Understanding the electronic transport mechanisms in molecular junctions is of paramount importance to design molecular devices and circuits. In particular, the role of the different junction components contributing to the current decay—namely the attenuation factor—is yet to be clarified. In this short review, we discuss the main theoretical approaches to tackle this question in the non-resonant tunneling regime. We illustrate our purpose through standard symmetric junctions and through recent studies on hybrid molecular junctions using graphene electrodes. In each case, we highlight the contribution from the anchoring groups, the molecular backbone and the electrodes, respectively. In this respect, we consider different anchoring groups and asymmetric junctions. In light of these results, we discuss some perspectives to describe accurately the attenuation factors in molecular electronics.

Keywords: molecular junctions; attenuation factor; density functional theory; graphene

---

1. Introduction

One of the main goals of molecular electronics is to mimic standard electronic circuits using molecules instead of p-n junctions like components based on silicon [1]. To do so, the first property to achieve in a molecular junction is to favor and understand at the fundamental level the circulation of the electronic current through the molecule. The problem of the electronic conduction mechanism in a molecule connected to metallic electrodes is very complex, and the possibility of many different regimes has been well presented theoretically by Reed et al. [2,3]. In particular, according to the molecular length, different regimes are observed, which exhibit different dependences in voltage and temperature. Hence, in the frame of elastic transport (considering that inelastic interactions only occur in the electrodes), for big molecular chains (longer than 5 nm), the electronic transport lies in an activated regime, called hopping regime, which is thermally activated. Indeed, in this regime, the conductance evolution of the molecular junction can be written as $G \sim \exp(-E_A/k_BT)$, where E_A represents the hopping activation energy, around 0.5 eV, k_B is the Boltzmann constant and T the temperature [4]. With respect to the molecular length, the conductance decreases linearly, which is easily understandable since the electrons have to jump (hop) from one molecular site to the nearest neighbor one [5,6]. This behavior is characteristic of Ohm's law, which we can observe at the macroscopic scale.

For smaller molecular chains, (i.e., below 5 nm) [7], the regime is not activated anymore and corresponds to the direct tunneling of the electrons through the molecular junction, provided that the applied voltage is lower than the characteristic electronic barrier of the system. In this case, there is no temperature dependence, and it is a reasonable approximation to say that the current varies linearly with the voltage at low bias. Indeed, the current–voltage relation is often non-linear for a significant range of voltages, before the molecular level is brought into alignment with the Fermi level of the electrodes. The transition between both regimes has been observed around 4 nm in conjugated polymers [8]. Note that, for larger voltages, this regime is generalized to the Fowler–Nordheim tunneling, where the current varies like the inverse of the voltage [9].

A fundamental problem in electronic transport in molecules lies in the length dependence of the conductance in the molecular junction. This dependence is reflected in the so-called attenuation factor, which is representative of the electronic current propagation in the molecular junction. Since a molecular junction is generally not metallic, the electronic conductance decreases as a function of the molecular length. Obviously, bearing in mind that the main application of molecular electronics is to design new devices for future electronics, one is interested in the lowest possible attenuation factor in order to increase the current in the molecular circuit. In the direct tunneling regime that we will particularly consider here, the conductance decays exponentially with respect to the molecular length, as we will detail in the next section. The main goal is therefore to reduce this exponential decrease as much as possible in order to optimize the electronic flow in the molecular circuit.

Hence, many experimental studies have been devoted to the determination of attenuation factors in different types of molecular junctions. In the meantime, theoretical methods have been developed to determine the electronic and transport properties of molecular junctions. The aim of this short review is not to describe extensively what has been done in the field, which would lead to an unreasonable amount of references, but more to discuss, in light of some representative systems, the progress and perspectives in theoretical methods to characterize the electronic transport in molecular junctions and to determine the attenuation factors.

In this respect, this review is organized as follows: in the first section, I will present a short state of the art of experimental determinations of attenuation factors in molecular junctions, and I will stress the most important results. Then, I will discuss the commonly used theoretical approaches, pointing out the corresponding strengths and weaknesses. In the fourth section, I will illustrate this discussion with the standard case of alkane-based molecular junctions, considering first the role of the anchoring groups in a symmetric junction with metallic electrodes and then by breaking this symmetry by using either different electrodes or different anchoring groups at each molecule sides. Obviously, since the anchoring groups are present in all the systems considered here, their influence will be analyzed in coordination with the different electrodes, molecular backbones and symmetry breakings. The underlying physical mechanisms will be addressed in light of these non-symmetric junctions. To complete this section, I will also discuss the role of the molecular backbone and its potential influence on the attenuation factor. Finally, I will summarize and draw some conclusions in the last section.

2. Some Attenuation Factors of Standard Molecular Junctions

In this review, we will consider specifically the non-resonant tunneling regime, which deals with small molecular lengths in the junction (typically below 5 nm [7]) and also low applied bias. In this case, the evolution of the conductance with respect to the molecular length can be written in the following form:

$$G = A \exp(-\beta L), \tag{1}$$

where L is the molecular length, A is the pre-exponential factor representative of the contact resistance at the molecule–metal interface, and β is the attenuation factor [10]. This attenuation factor is representative of the current attenuation within the molecular junction, with respect to the molecular length. In order to redefine the context, the attenuation factor is zero for a pure metallic junction, as no attenuation occurs, whereas it is of the order of 25 nm^{-1} for the vacuum where the current is fully attenuated.

A standard example of attenuation factor measurement is alkane-based molecular junctions. Many studies have been devoted to this system, and the attenuation factor currently lies around 7–9 nm^{-1} [10–12]. It has to be noted that this value is strongly dependent on the chemical nature of the molecular wire, since substituting carbon by silicon to constitute what is called oligosilane chains reduces the attenuation factor to 3.9 nm^{-1} [13]. This effect is even more enhanced considering germanium-based molecular wire, to form a germane chain, leading to $\beta \sim 3.6$ nm^{-1} [14]. In this respect, we can speculate that going down the lines of the periodic table, i.e., increasing the atomic number of the chain constituents, would lead to an important reduction in the attenuation factor—in other words, a higher conductance, potentially related to the increasing number of electrons per atom. On the other hand, still considering non-conjugated molecular chains, the opposite behavior is found in siloxane chains, where the repetition unit is based on a Si-O dimer, with a very important attenuation factor of β ~ 12.3 nm^{-1} [15]—in other words, almost no conductance.

Beyond the chemical nature of the molecular chain elements, the nature of the molecular bonding plays an important role. For example, coming back to carbon elements, aromatic chains present reduced attenuation with respect to alkane chains, with β ~ 2.5 nm^{-1} [16]. Similarly, carotenoid polyenes present β of 2.2 nm^{-1} [17] and oligothiophenes have β between 2 and 3 nm^{-1} [18]. Consequently, one can deduce that the aromaticity of the molecular chain helps in reducing the attenuation factor; however, it is more or less always in the same range.

Besides the above, some more complex molecular systems also exhibit even lower attenuation factors, around 10 times lower than the polyenes discussed previously. For example, polymethine dyes, which are π-conjugated compounds with an odd number of carbons, have attenuation factor β ~ 0.4 nm^{-1} [19]. This is mainly attributed to greater electronic delocalization, in comparison to standard polyenes, enhanced by a smaller degree of bond order alternation. However, this behavior seems to be limited by the molecular length, leading to a more resistive junction for long molecular chains. In addition, the more exotic case of porphyrin polymers is very interesting. Indeed, when considering oligo-porphyrin molecular junctions, based on Zn porphyrins with pyridine ligands, one can observe a similar reduced attenuation factor β ~ 0.4 nm^{-1} [20]. Actually, it seems that the molecular conductance has a strong temperature dependence and a weak length dependence, even though it does not correspond to the hopping mechanism described in the introduction. It is rather consistent with phase-coherent tunneling through the whole molecular junction. Even more surprising is the result obtained by Leary et al., who have studied molecular chains of fused porphyrins [21]. Namely, in the considered porphyrin oligomers, the monomers connect directly with their nearest neighbor through the porphyrin cycle. In this case, contrary to the previous cases discussed here, the conductance increases with the distance, by more than a factor of 10, when a small bias (~0.7 V) is applied to the junction. This exceptional behavior is due to the evolution of the HOMO-LUMO gap that rapidly decreases with the molecular length, which compensates for the increased tunneling distance. Finally, the last example to be cited with respect to low attenuation factors is the work conducted by Brooke et al., who show a structural control on the electronic transport resonance in HS(CH$_2$)n[1,4 $-$C$_6$H$_4$](CH$_2$)$_n$SH (n = 1, 3, 4, 6) metal–molecule–metal junctions, leading to very small attenuation factors [22]. This work offers very promising perspectives in gating the transport resonance in order to modulate the molecular junction behavior.

In this discussion, we have mainly considered the case of single molecule junctions. An interesting question that arises is this: what happens if we consider not only a single molecule junction but also a large molecular area? In this case, several experiments have found that the attenuation factor β mainly remains the same, and only the prefactor of the conductance is modified, as for molecular junctions based on biphenyl, nitrophenyl, ethynyl benzene, anthraquinone, etc., where one can observe attenuation factors around 2.7 nm^{-1} [23] (see Figure 1).

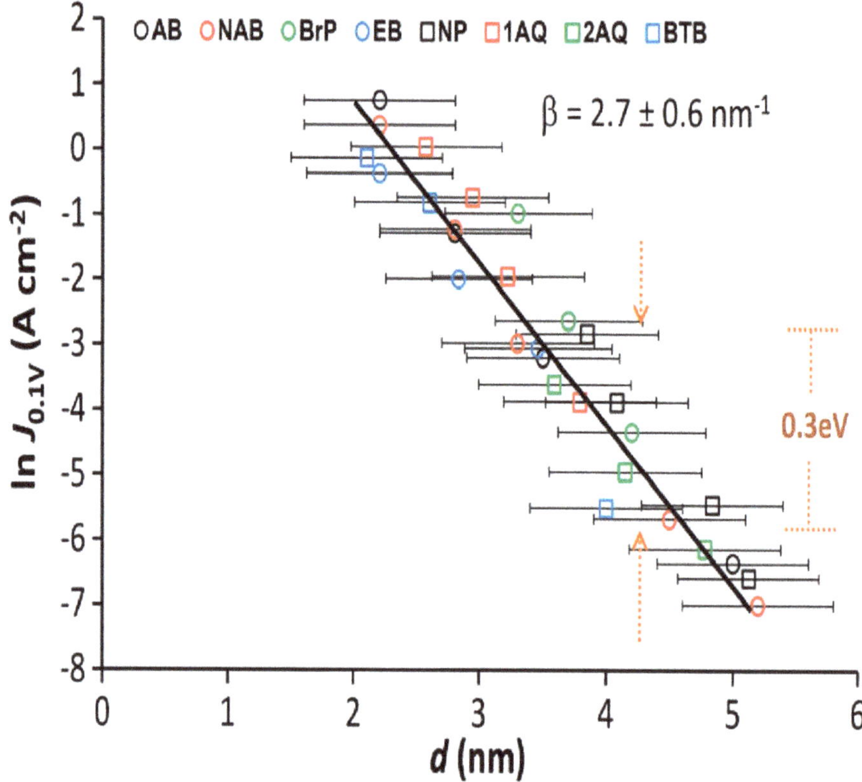

Figure 1. Overlay of attenuation plots for eight different molecules constructed from J–V curves with different thicknesses of each structure (the length of the error bars is two standard deviations). The lines are least squares regression lines for aromatic (2.7 nm^{-1}) molecules. The different abbreviations stand for the aromatic molecular names: NP = nitrophenyl, AQ = anthraquinone, NAB = nitroazobenzene, BrB = bromophenyl, EB = ethynylbenzene, AB = azobenzene and BTB = bisthienylbenzene. Adapted figure from [23] and with permission. Copyright (2010) American Chemical Society.

As an important consequence, these results show that the study of single molecule junctions is not only important at the fundamental level. The understanding of the transport mechanism at the single molecule level and, in particular, the determination of the attenuation factor may be extrapolated to large molecular areas, which are currently used in molecular electronics devices. Another important aspect that can be stressed is that, below 5 nm, it seems that there is no significant difference in the attenuation factor with respect to the molecular backbone of the aromatic molecules considered in the junction. This might be attributed to the relative positions of the molecular electronic levels, which remain constantly pinned near the Fermi level at around 1.3 eV [23,24]. We will see in the next sections that this position of the molecular levels is an important parameter to determine the attenuation factor. However, it has to be noted also that most of these experimental results have been obtained using molecular junctions in symmetric configurations, namely with the same electrode at each side of the junction (most of the time, metallic electrodes made of noble metals) and similar anchoring groups. Very few studies have considered non-symmetric junctions—for example, with different anchoring groups at each molecular end—which might have an influence on the position of the molecular levels and yield other kinds of information about the transport mechanism [25,26]. In addition, as we will detail later, some recent experiments have started to use different electrodes at each end as well as non-metallic electrodes. Obviously, the question that arises is this: what is the fundamental element in a molecular junction that determines the attenuation factor? Is it the chemical nature of the molecular backbone? Is it the anchoring groups? Is it the electrode or the coupling to the electrode? Alternatively, is it a mix of all these aspects? To answer these questions, theoretical modeling of the molecular junction and of the electronic transport through the junction is of paramount importance. In particular, it should help to discriminate these different contributions. In the next section, we will consider the most common theoretical approaches to characterize the electronic transport in a molecular junction and to determine the attenuation factor.

3. Theoretical Approaches

In this section, we will discuss the most common theoretical approaches used to determine the electronic transport and in particular the attenuation factor in a molecular junction. An extensive review of electronic transport calculations using ab initio and density functional theory (DFT) based methods and Green functions can be found in [27]. Unfortunately, a direct relationship with the attenuation factor and overall a deep interpretation is not necessarily straightforward to deduce from these calculations. The objective here is to present the main contribution from the different parts of the molecular junctions that are characterized theoretically in each approach.

Hence, the most common description of the electronic current in a molecular junction has been shown by Simmons [1,28], considering the tunnel effect between metallic electrodes and a thin insulating film. In this respect, the current and the electronic conductance follow an exponential decay, as proposed in Equation (1), where the attenuation factor β can be expressed as $\beta \sim \sqrt{(2m\varphi/\hbar)}$, with m the mass of the electron, $\hbar$ the reduced Planck constant and where φ represents the electronic potential barrier of the system. In a molecular junction, this barrier is nothing other than the energy difference between the closest (non-resonant) molecular level to the Fermi level of the electrode and the Fermi level [1,10]. Namely, if we consider electron transport, β will depend on the $E_{LUMO}-E_{Fermi}$ difference, whereas, if we consider hole transport, β will depend on the $E_{Fermi}-E_{HOMO}$ (E_{LUMO}, E_{HOMO} and E_{Fermi} being, respectively, the energy position of the LUMO, HOMO and Fermi level of the system). The immediate consequence of this interface property is that, in principle, beyond the nature of the molecular backbone, one could modulate the attenuation factor by changing the specific anchoring groups forming the bond between the molecule and the electrode. Indeed, the position of the molecular levels with respect to the Fermi level is mainly influenced by the anchoring groups used to connect the molecule to the electrodes. This aspect will be detailed later when considering properly the influence of the anchoring groups in well-established examples. However, this heuristic model is inspired by scanning tunneling microscope (STM) experiments, or the standard evolution of a wave function through a potential barrier in quantum mechanics, and represents the simplest expression of the Simmons model [28], currently used in molecular electronics to describe conductance attenuation in molecular systems. It works rather well for molecular junctions where the levels are close to the Fermi level but requires more ingredients when this is not the case.

Other more sophisticated models highlight the role of the molecular backbones without taking into account the interfaces between the molecule and the electrode. For example, from an atomistic point of view, the conductance of the molecular wire can be understood using a simple tight-binding model, with N sites of energy $\varepsilon_1, \varepsilon_2 \ldots \varepsilon_N$ coupled through hoppings $t_{1,2}, t_{2,3}, \ldots . t_{N-1,N}$ and bridging the two electrodes [29] (see Figure 2).

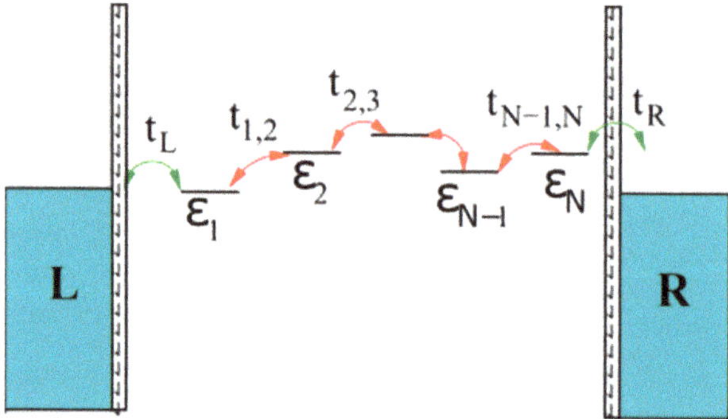

Figure 2. Schematic representation of the bridge model with N sites of energy $\varepsilon_1, \varepsilon_2 \ldots \varepsilon_N$ coupled through hoppings $t_{1,2}, t_{2,3}, \ldots . t_{N-1,N}$ and bridging the left and right electrodes. Reprinted figure from [1] and with permission. Copyright (2010) World Scientific.

Notice that these energy sites are independent of the coupling at the interfaces, which automatically eliminates the role of the anchoring groups. Moreover, we can consider that all these energies are identical, namely to ε_0, and the same for the hoppings, which are all equal to t. In this approach, the attenuation factor can be expressed as:

$$\beta(E) = (2/a) \ln |(E - \varepsilon_0)/t|, \tag{2}$$

where a measures the segment size, for a total molecular length of Na [1]. Considering the energy E as the Fermi level, and average values of $|(E - \varepsilon_0)/t| = 10$ and a = 5 Å, this expression gives typical values of β of around 9 nm^{-1}, independently of the anchoring groups or the electrodes. Note that a similar result can be deduced from the dispersion relation of an infinite linear chain:

$$2t \cosh(\kappa a) = E_{Fermi} - \varepsilon_0, \tag{3}$$

with $\kappa = \beta/2$, $(E_{Fermi} - \varepsilon_0)/t >> 1$, and where the hyperbolic cosine stands for the exponential behavior at each side of the molecular chain where the electronic wavefunction tunnels from or to the electrode. In other words, this approach is also a generalization of the Simmons model. However, here, the energy difference $E_{Fermi} - \varepsilon_0$ corresponds to the potential barrier of an infinite molecular chain without anchoring groups or electrodes, namely ε_0 being an orbital of the infinite molecular backbone and not, for example, the HOMO level of the junction.

Finally, the last important model lies in the determination of the attenuation factor through a complex bandstructure calculation [30,31]. Similarly to what happens in solid state physics, an infinite molecular chain (analog to an infinite crystal) is considered and its bandstructure is calculated. Real wavevectors correspond to the usual molecular spectrum of the junction.

The attenuation factor is determined from the calculation of evanescent states of the electronic wavefunction inside the molecule, namely the electronic states corresponding to complex wavevectors. These evanescent states, which correspond to the Schockley surface states in solid crystals [32], correspond to the different potential conduction channels of the molecular junction, originating from the different orbital hybridizations. Obviously, among all these different channels, the current will follow the channel with the lowest attenuation factor, similarly to the macroscopic behavior. Therefore, the value of the attenuation factor can be read on the complex bandstructure for the evanescent state with the smallest extension in the gap, as represented in Figure 3. This approach also leads to attenuation factors around 8 nm^{-1} for alkane chains. Notice that an interesting mathematical derivation establishes a link between this result and a square root variation for the attenuation factor similar to the one of the Simmons model. However, this expression is valid again only for the infinite molecule and not for the one whose HOMO and LUMO levels are determined by the coupling at the interface.

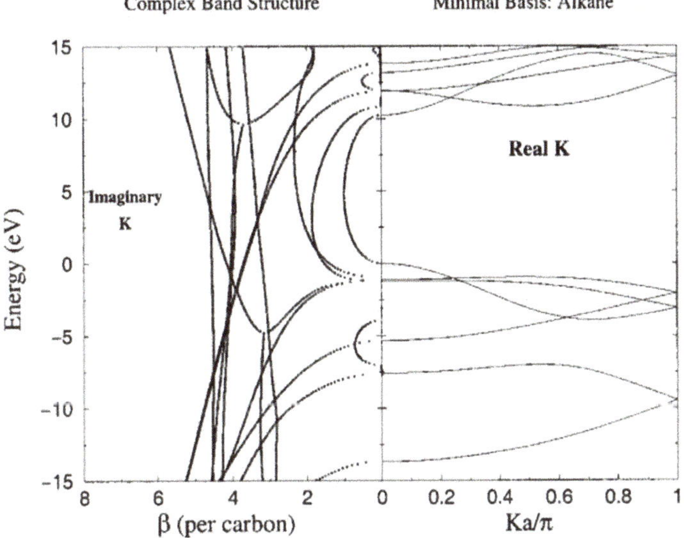

Figure 3. Calculated complex bandstructure for an infinite alkane molecular chain. The left and right part of the graph correspond to the imaginary and real wavevectors, respectively. The attenuation factor can be read on the horizontal axis, in the imaginary wavevector part, considering the localized state with the smallest extension (semielliptical-like curve). Reprinted figure from [30] and with permission. Copyright (2002) by the American Physical Society.

To summarize the different theoretical approaches presented in this section, we can observe that each approach stresses a specific contribution. Hence, the model considers either the anchoring groups/electrodes (namely the coupling of the molecule at the interface) through the relative position of the molecular levels with respect to the Fermi level or the molecular backbone (namely the chemical nature of the molecular chain), in a rather exclusive manner. Consequently, it remains difficult to determine accurately and for all kinds of molecular junctions the corresponding attenuation factor. As such, a universal model does not seem to exist yet. Therefore, the debate remains open regarding whether the attenuation factor is determined by interface effects or only by the molecular backbone or by a combination of both. A large experimental consensus seems however to be established in favor of a major influence of the molecular backbone. As a remark, we can stress nevertheless that all of these theoretical studies (as already discussed in the previous section for most of the experimental studies) have only considered symmetric cases, where electrodes and anchoring groups at each side were similar. The study of a non-symmetric case through these models would be of high interest to test their range of applicability.

In the next section, we will illustrate these different models with standard test cases, namely alkane molecules sandwiched between metallic electrodes, and we will investigate the respective roles of the electrodes, the anchoring groups and the molecular backbone, by investigating asymmetries in the junctions.

4. A Test Case: Alkane Chains

4.1. Role of the Anchoring Groups: Symmetric Junctions

Alkane polymers probably constitute the most common and the simplest molecular chains that have been studied in molecular electronics. Indeed, from their simple chemical nature, their easy availability and ability to connect to different chemical groups or metals and despite their very large gap, which would in principle reduce the conductance, it represents a toy model for electrical conduction in molecular systems. Hence, it is no surprise that this system has been measured in numerous works using different metallic electrodes (mainly noble metals Au, Ag, Cu) and different anchoring groups. Here, we briefly present the well-documented results in the literature on symmetric alkane-based molecular junctions, before exploring further the underlying physical mechanisms through the study of hybrid junctions. For example, anchoring groups like thiol (-SH) [33], amine (-NH$_2$) [34], carboxylic acid (-COOH) [35], isocyanide (-NC) [36], methyl sulfide (-SMe) [37], etc., have been studied extensively. What is particularly interesting is the comparison of the measured attenuation factors for several junctions with different anchoring groups. For example, Chen et al. [38] performed such a comparison between thiol, amine and carboxylic acid anchoring groups. The first important difference in the respective electronic properties of these junctions lies in the contact resistance, which is inversely related to the prefactor A in Equation (1). Indeed, the contact resistance is smaller for thiol, bigger for amine and even bigger for carboxylic acid, which yields overall conductance higher for thiol than for amine and then higher for amine than for carboxylic acid. This is due obviously to the different kinds of electronic coupling at the interface, namely of covalent nature for the thiol, much weaker for the amine and related to a deprotonation process of the carboxylic acid to contact the electrode. In this respect, geometries at the interfaces are also affected and the involved symmetries are different, as we will see in more detail later. In addition, the electronic properties seem to be rather different, since the HOMO level is located at around 2.0 eV from the Fermi level in the thiol case, 5.5 eV for the amine case and 1.1 eV in the carboxylic acid case. These differences in HOMO level positions are attributed to different symmetry couplings at the interfaces, as will be discussed in the next section. However, despite these important differences, the attenuation factors seem to remain rather similar for the three junctions, around 0.8–0.9 per C atom (or –CH$_2$ unit). Considering such similarities in the attenuation factors for different anchoring groups, one can wonder what the real impact of the anchoring groups on the attenuation factor is. From these first results, the only incidence that we can deduce is a variation in the contact resistance and consequently a variation in the overall conductance. In the next section, we will consider the role of the electrode, using different anchoring groups, in the electronic transport in molecular junctions. The main idea is to compare the influence of the anchoring groups with the same or different electrodes.

4.2. Role of the Electrodes: Hybrid Junctions Using a Graphene Electrode

Most of the conductance measurements in molecular electronics have used metallic electrodes and, in particular, noble metals, either using mechanically controlled break junctions (MCBJ) [39], scanning tunneling microscopy (STM) [40], conductive probe atomic force microscopy (CP-AFM) [41] or also the STM-based I(s) method developed by Nichols et al. [42]. Nevertheless, an important number of recent works have been devoted to the use of carbon electrodes or even graphene substrates [43,44]. In this respect, we will discuss here the attenuation factors measured experimentally and determined theoretically on hybrid molecular junctions with a gold electrode at one molecular end and a graphene electrode at the other. We will also consider the role of different anchoring groups using this graphene electrode.

First, we start by considering alkanedithiol molecules of different lengths, probed experimentally using the I(s) method and modeled using DFT and electronic transport calculations to interpret the obtained results [45]. An atomic model of the system is represented in the left part of Figure 4 for a butanedithiol sandwiched between a gold and a graphene electrode, forming a hybrid metal/molecule/graphene junction. The conductance has been measured for different molecular lengths between 2 and 12 –CH$_2$ groups, allowing us to deduce the length dependence of the conductance. In this respect, the use of a graphene electrode does not affect the exponential conductance decay. In parallel, electronic transmission spectra and conductance have been calculated after DFT structural optimization of the junctions and density of states calculations. In addition, the contact resistance has been determined to be significantly larger than what is obtained for a standard symmetric junction with two gold electrodes. This is due to the weak coupling (of van der Waals nature) at the graphene–molecule interface.

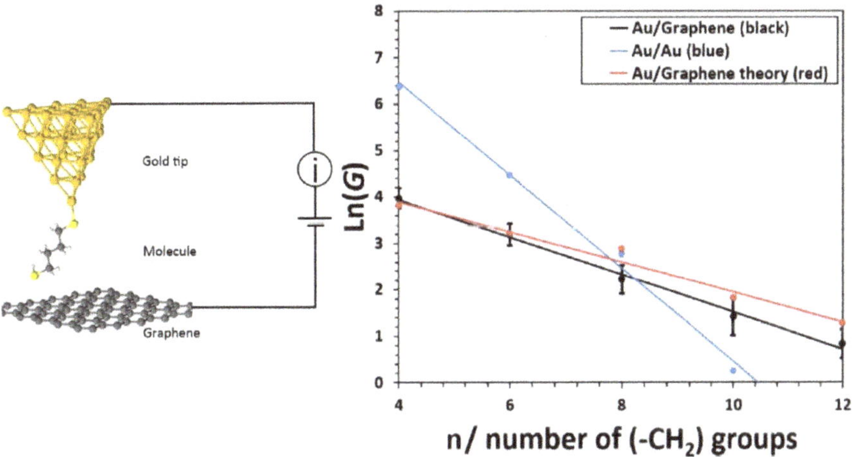

Figure 4. (**Left**) Schematic representation of the hybrid metal–molecule junction formed in the experiment for conductance measurements. (**Right**) Evolution of the conductance as a function of the molecular length: experimental measurements on asymmetric metal–graphene junction (black), experimental measurements on symmetric metal–metal junction (blue) and conductance calculation on asymmetric metal–graphene junction (red). Reprinted figure from [40] and with permission. Copyright (2016) American Chemical Society.

The length evolution of the experimental and theoretical conductance is represented on the right part of Figure 4.

As a result, the attenuation factor for this hybrid junction is approximately half of the one obtained for its symmetric metal–molecule–metal counterpart. Namely, the attenuation factor now lies at around 0.4 per carbon atom. Hence, even though the contact resistance is quite high for the hybrid junction due to the weakly coupled interface with graphene, for molecules longer than ~1 nm, the metal–molecule–graphene junction turns to be more conductive than the standard metal–molecule–metal one. Moreover, for similar molecular lengths, the hybrid junction is also longer than the standard one, due to the van der Waals contact which requires a 3 Å distance between the molecule and graphene. Therefore, the question arises: how can such unusual behavior occur? We will try to understand it in the frame of the Simmons model, the most common theoretical approach to determine an attenuation factor that we have discussed in Section 3. Hence, the electronic behavior can be explained as follows by the different couplings at the electrode–molecule interfaces and the different molecular level alignments. When a thiol-terminated alkane is adsorbed on a gold surface, due to the strong interface electrostatic dipole, there is an important charge transfer, which partially depopulates the molecule and moves the HOMO level toward the Fermi level of the surface. If the same molecule is contacted between two gold electrodes, there is an interface dipole at each molecular extremity, in opposite directions, leading to a cancellation of the two dipoles. Then, the HOMO level remains far from the Fermi level, at around 2 eV, as discussed previously, yielding an important electronic barrier that is reflected in a high attenuation factor ($\beta \sim 8.6$ nm-1 or 0.8 per C atom [1,11]). Note that Brooke et al. have also studied the case of symmetric junctions, where it was argued that dipoles in symmetric junctions can result in the HOMO level being dragged down in energy [22]. If the second electrode is now a graphene plane, the interface dipole at the graphene side ruled by van der Waals interaction is much weaker than the one at the gold side, which is not cancelled in this situation. This results in an important charge transfer at the gold–molecule interface and a shift of the HOMO level toward the Fermi level. Consequently, the energy difference E_{Fermi}–E_{HOMO} is reduced to 0.4 eV, which reduces considerably the attenuation factor, to 0.4 per C atom, as observed experimentally. In this case, as we can see, the simple Simmons model is perfectly able to describe the electronic behavior of the junction. Moreover, we have observed that the use of a graphene electrode at one extremity breaks the electrostatic symmetry of the system, leading to a reduced attenuation factor for thiol groups, as compared to the same junction and two gold electrodes.

Now, in order to extend the comparison to symmetric junctions, and to probe the theoretical model used for the thiol case, we will consider another hybrid junction with different anchoring groups, namely amine groups [46]. The same procedure has been applied here with I(s) measurements and DFT based electronic transport calculations. The corresponding results are presented in Figure 5. First, we consider the evolution of the conductance as a function of the molecular length, in a logarithmic scale. Similarly to the case of the thiol Au–graphene junction, an attenuation factor of 0.4 per C atom is also found in the amine case. This behavior is consistent with what happens with Au–Au junctions, as we have seen in the previous subsection. Indeed, for all the anchoring groups, the attenuation factor remains practically the same. Here, the introduction of a graphene electrode seems to result in a similar effect, but it is reduced to around half its value with respect to Au–Au junctions. What is more surprising here is the direct comparison of the conductance with the Au–Au junction and the same amine groups. Indeed, while, for the thiol case, we observed greater conductance above a certain molecular length due to the reduced attenuation factor, we can observe for the amine group a more important conductance independently of the molecular length is, as shown in Figure 5. Even more surprisingly, while, for the Au–Au junction, the thiol case presents higher conductance than the amine one (due to better contact and lower contact resistance), in the Au–graphene junction, the amine case presents higher conductance than the thiol one.

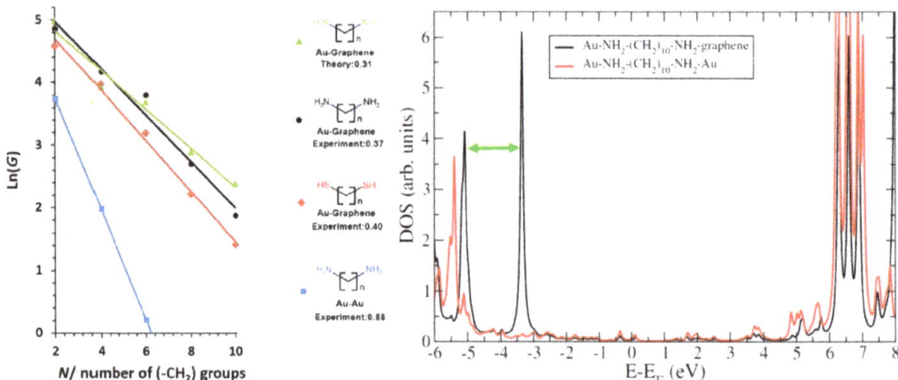

Figure 5. (**Left**) Evolution of the conductance as a function of the molecular length for molecular junctions with different anchoring groups and electrodes: conductance calculation on Au–graphene junctions with amine group (green), conductance measurements on Au–graphene junctions with amine group (black), conductance measurements on Au–graphene junctions with thiol group (red) and conductance measurements on Au–Au junctions with amine group (blue). (**Right**) Density functional

theory (DFT) calculated density of states (DOS) for Au–graphene and Au–Au molecular junctions with amine anchoring groups. Reprinted figure from [46] and with permission. Copyright (2017) American Chemical Society.

In order to understand this particular behavior, we have calculated the density of states (DOS) for the alkane-based junction with amine anchoring groups, using Au–Au or Au–graphene electrodes. The result is represented in the right part of Figure 5.

The calculated DOS allows us to determine the position of the HOMO level (which is the closest level to the Fermi level and therefore the conductive one in the junction, as for the thiol junction), so that we can apply the Simmons model used previously to determine the attenuation factor and the overall conductance. However, the calculated DOS indicates here a HOMO level located not around 0.4 eV below the Fermi level but 3.5 eV below the Fermi level for the amine anchoring group. As a consequence, the use of such a big potential barrier in the Simmons model would necessarily lead to a huge attenuation factor, which is not in agreement with the experimental observations. Therefore, this simple example highlights some limitations of the models presented in Section 3. Indeed, other models based on complex bandstructure calculations would not help either in determining correctly the attenuation factor, as electrode asymmetry cannot be taken into account. Note that the determination of the attenuation factor for the symmetric Au–Au junction is complicated as well, since, in that case, the energy barrier is about 5.5 eV, which does not fit into the Simmons model either. From this system, we can deduce that there is no universal model to calculate accurately the attenuation factors in molecular junctions. Hence, the Simmons model can be used only for specific ranges of electronic potential barriers, and the other models seem to be valid only for symmetric junctions, without properly considering the influence of the anchoring groups or the electrodes.

In [46], this problem has been solved by considering that, between the Au–Au and the Au–graphene amine junctions, the HOMO level has been relocated from the same amount as in the Au–Au and the Au–graphene thiol junctions, namely around 2 eV (as indicated by the green arrow on Figure 5). Since the attenuation factors are the same for both anchoring groups in Au–Au junctions, it can be deduced that they should be the same for the Au–graphene junctions. The underlying reason is the fact that the coupling to the electrode and the molecular conduction are different for thiol and amine anchoring groups. The thiol groups connect very well to the electrodes, as of π symmetry, which explains the small electronic barrier, whereas the amine groups connect poorly to the electrodes, as of σ symmetry. Conversely, the thiol groups connect poorly to the alkane molecular backbone, yielding low intramolecular electronic propagation, whereas the amine groups connect much better, which increases the intramolecular propagation. One effect compensating the other, both anchoring groups lead to the same attenuation factor, mainly depending on the electrodes. This balance between coupling to the molecular backbone, leading to a good intramolecular conductance, and coupling to the electrode, which reduces the interface potential energy barrier, also explains the similar attenuation factors for different anchoring groups in alkane-based symmetric junctions between gold electrodes.

One last point has to be clarified regarding the overall conductance of the amine junction, and the explanation is found again in the DOS calculation. Indeed, the HOMO level is different in the case of Au–graphene electrodes, due to a level splitting with respect to the Au–Au junction. Indeed, the introduction of the graphene electrode breaks the symmetry of the molecular junction (this effect was also observed for the thiol junction, where the symmetry breaking was seen in the non-compensation of the electric dipoles) and splits the original HOMO level, leading to a reduced molecular gap (by about 2 eV). Consequently, since the overall conductance depends on the self-energies that couple the molecule to the electrodes, and these self-energies vary as the inverse of the molecular gap, this important reduction of the gap leads to a much higher molecular conductance. This is why the Au–graphene alkanediamine junction presents more important conductance than its Au–Au counterpart but also than the alkanedithiol junction, where the introduction of the graphene electrode shifts the molecular levels without any gap reduction.

4.3. The Case of Platinum–Graphene Hybrid Junctions

In the previous subsection, we have considered hybrid Au–graphene electrodes for two different anchoring groups in alkane-based molecular junctions and we have observed an important reduction in the attenuation factor. Here, we will observe the effect of a change in the metallic electrode by substituting the gold electrode with a platinum one. As a matter of comparison, we consider again the alkanedithiol molecular junction, following the same experimental and theoretical procedure [47]. Similarly to the Au–graphene alkanedithiol molecular junction, the conductance presents an exponential decay, still within the framework of the non-resonant tunneling. The corresponding attenuation factor is slightly lower, around 0.3 per C atom, which is attributed mainly to the difference in work function between gold and platinum. It has to be noticed that this system is again very well modeled through the Simmons approach, because of the strong coupling to the electrode and the reduced electronic potential barrier.

Moreover, we can deduce from this result, combined with the results obtained on Au–graphene junctions, that the reduction of the attenuation factor is caused by the symmetry breaking of the graphene introduction in the system. This is not related to the nature of metallic electrode, but rather to the difference of interaction strength at each electrode interface. Indeed, the metallic electrode is coupled covalently to the molecule through the anchoring group, while the graphene electrode is coupled in a much weaker manner through van der Waals interaction. This is particularly true in the case of the thiol anchoring group, which deprotonates at the gold interface, leading to a –S radical (thiolate) that is very reactive with the gold electrode, whereas it remains in the thiol form –SH at the graphene interface, leading to van der Waals contact. Consequently, this is the interaction symmetry breaking at each molecular end, covalent/van der Waals, which leads to the important decrease in the attenuation factor and the increase in the molecular conductance. In this respect, one can anticipate that a molecule weakly coupled to two graphene electrodes through van der Waals interaction would probably present a similar attenuation factor as the one obtained for symmetric Au–Au junctions. As a remark, a similar junction has been studied using a graphitic tip for the I(s) measurements and the usual graphene electrode. The resulting attenuation factor was found to be very similar to the one of the Au–graphene junction, around 0.4 per C atom. This was due again to a coupling difference, since the graphitic tip was very reactive in this situation and coupled covalently to the molecule, whereas the graphene counter electrode coupled weakly to the molecule [48].

4.4. Role of the Anchoring Groups: Asymmetric Junctions

As we know now, symmetry effects are very important in electronic transport in molecular junctions. In particular, we have seen that it has a strong influence on the attenuation factor. For this reason, it is interesting also to see what would be the influence of considering different anchoring groups at each molecular side to connect the electrodes to the molecules. In this respect, we highlight here the interesting case of hybrid Au–S–alkane–COOH–graphene molecular junctions, where thiol and carboxylic acid groups have been used at each extremity of the junction with gold and graphene electrodes [49]. An atomic representation of the molecular junctions studied here is represented in Figure 6.

Obviously, from the theoretical considerations discussed above, there was no valid approach until now that was able to describe accurately such an asymmetric junction. Moreover, this type of junction has also been studied previously using Au–Au electrodes. As a reminder, the attenuation factor of junctions using only thiol or only carboxylic acid as anchoring groups are very similar, around 0.8–0.9 per C atom. Then, the corresponding attenuation factor for an asymmetric junction using different anchoring groups was found to be of the same magnitude, so without important change with respect to the symmetric junctions [25]. When considering junctions with Au–graphene electrodes, the situation is slightly different. Indeed, for a symmetric junction with carboxylic acid at both ends, the attenuation factor is around 0.7 per carbon atom, probably due to the low coupling of the carboxylic acid to the electrodes, which does not make a lot of difference in the situation of two gold electrodes. However, the situation is slightly different in the Au–graphene case, where the attenuation factor is around 0.4 per carbon atom, namely the same value as that found previously for thiol and amine groups in Au–graphene junctions. Interestingly, the electronic behavior is rather different. In thiol overall, but also in amine molecular junctions, the electronic transport is achieved through the HOMO level which is the closest level to the Fermi level. Here, as shown in Figure 7, the Fermi level is rather located near the middle of the HOMO and LUMO gap of the molecule, similarly to what is observed in the same junction with carboxylic acid at each molecular end. In terms of electronic transport, it means that there is no dominant molecular level (HOMO or LUMO) to the molecular conductance.

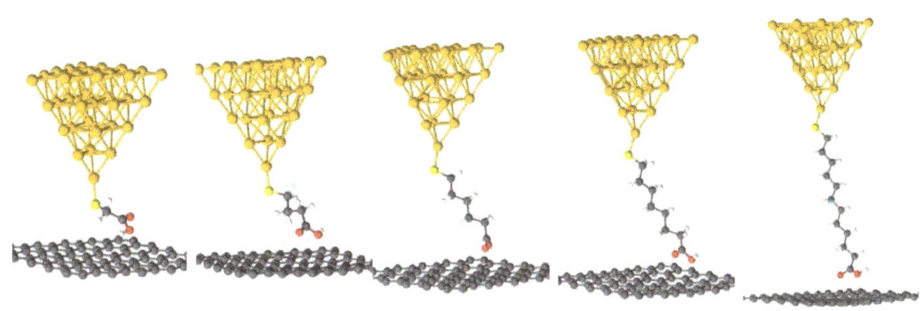

Figure 6. Atomic representation of the different hybrid Au–graphene alkane-based molecular junctions using thiolate and carboxylic acid anchoring groups at each extremity. Reprinted figure from [49] and with permission. Copyright (2019) Wiley-VCH Verlag GmbH & Co.

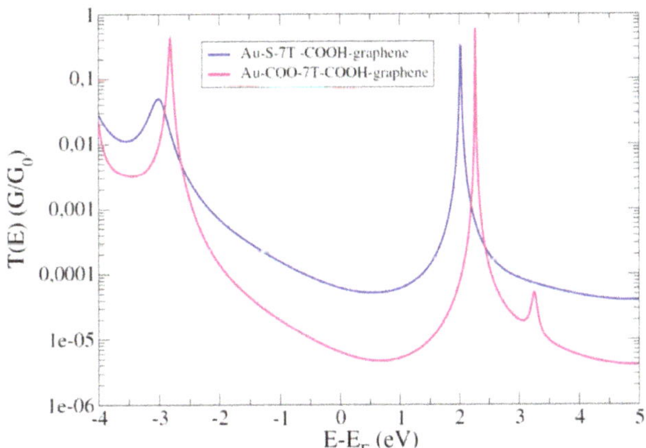

Figure 7. DFT calculated electronic transmissions for Au–graphene molecular junctions with carboxylic acid anchoring groups at each extremity (pink) and a thiol group at one extremity and a carboxylic acid at the other one (blue). Reprinted figure from [49] and with permission. Copyright (2019) Wiley-VCH Verlag GmbH & Co.

From this result, we can deduce that molecular junctions with asymmetries in the anchoring groups see their electronic transport driven by the anchoring group that couples the most, even though the electronic structure is also strongly affected by the less coupled anchoring group. In other words, the current flows along the most conductive channel, similarly to what happens at the macroscopic scale, namely the channel which presents the best coupling at the molecule–electrode interface or the lowest contact resistance. Certainly, this constitutes only one particular study of asymmetric junctions in terms of anchoring groups and many other examples should be probed to extract trends that are more general. Nevertheless, it appears rather intuitive that the most conductive anchoring group will favor the most conductive channel and therefore the lowest attenuation factor.

5. Importance of the Molecular Backbone: Conjugated Molecular Wires

In the previous section, we have focused the discussion on alkane molecular chains and we have considered the different parameters which can influence the electronic transport in the junction and consequently the attenuation factor. In particular, we have considered the anchoring groups, in symmetric or asymmetric junctions, and the electrodes, with different metals and with a graphitic tip. Without being exhaustive, it is important also to consider what happens when using another molecular backbone, like conjugated molecular wires. To this end, we chose to have a short look at polyphenylene chains, which are also very common and popular for molecular electronics. These polymers are very interesting, since, as presented in Figure 1 and [15], their attenuation factors are around three times smaller than for the alkane chains, making these molecular wires much more conductive. Hence, with what we have seen by using graphene electrodes, we can expect an even higher conductance for these polymers used in hybrid molecular junctions.

Therefore, hybrid Au–graphene and standard Au–Au molecular junctions have been studied through I(s) measurements and DFT calculations, with polyphenylenes, using thiol and amine anchoring groups, as a direct comparison with alkane-based molecular junctions [50].

Surprisingly, the electronic behavior is very different from what was expected. Indeed, either experimentally or theoretically, almost no difference has been found for the attenuation factors determined for Au–Au and Au–graphene junctions. Moreover, the different anchoring groups, thiol and amine, did not bring any significant difference. Again, this can be better understood from the DOS calculations represented in Figure 8. The projected DOS on the molecular part (including the anchoring group) reveals that there is almost no difference in the electronic structure of each molecular junction. This means that, in this case, the use of a graphene electrode does not break the symmetry, as was the case for the alkane chains. This is further illustrated in Figure 8 with the representation of the spatial extension of the HOMO level for both Au–Au and Au–graphene junctions. In both cases, we can observe a molecular orbital of π symmetry which propagates well along the molecular chain. However, we can also observe that the coupling to the electrodes remains the same in both junctions, despite the introduction of the graphene electrode. In other words, for this system, there is no symmetry breaking induced by the introduction of the graphene electrode. Consequently, the coupling to the electrode or the anchoring group used will have no effect on the electronic structure of the system and therefore on the electronic transport. Hence, the polyphenylene molecular chain solely drives the attenuation factor. From a theoretical point of view, in this case, the Simmons model would not help in discriminating the effect of the anchoring group, whereas a complex bandstructure calculation would probably give the correct attenuation factor, independently of the electrodes or the anchoring groups.

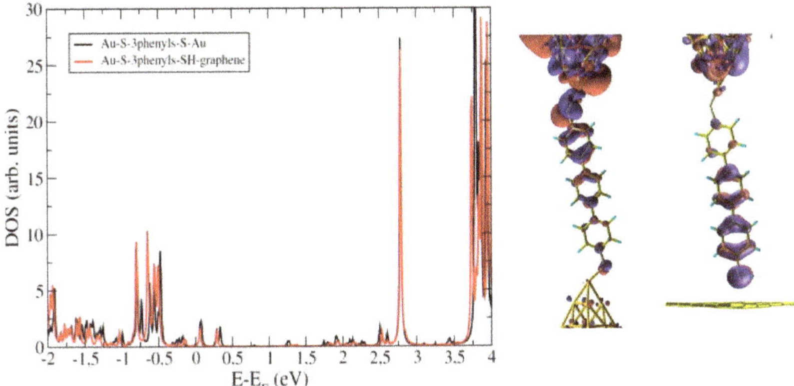

Figure 8. (**Left**) DFT calculated projected density of states for Au–graphene and Au–Au triphenyl molecular junctions using thiol groups. (**Right**) Representation of the corresponding HOMO wavefunction for both junctions. Reprinted figure from [50] and with permission. Copyright (2019) American Chemical Society.

6. Summary

In this review, we have introduced the main electronic transport mechanisms in molecular junctions, to discuss mainly the non-resonant tunneling regime. In particular, we have considered as a characteristic of this regime the attenuation factor in the current exponential decay, observed experimentally and determined theoretically. In this respect, we have discussed the main theoretical approaches proposed to date to describe the attenuation factors. As we have seen, there is unfortunately no universal model to characterize the electronic transport in a molecular junction. Beyond performant DFT calculations, which allow us to calculate the electronic transmission and to deduce the attenuation factor, a full understanding of the electronic transport in a molecular junction remains complicated. Indeed, each present theoretical approach describes either the role of the anchoring group, by considering an accurate evaluation of the molecular levels position with respect to the Fermi level, or the role of the molecular backbone, without considering properly the coupling to the electrode or the interface states related to the anchoring groups. In addition, most of the studies consider symmetric junctions with the same electrodes and same anchoring groups at each molecular side.

Then, we have illustrated these aspects by shortly reviewing experimental and theoretical studies on alkane-based molecular junctions. Starting with standard junctions with gold electrodes at each end, we have also considered a hybrid junction where one gold electrode is substituted by a graphene electrode. In this case, and for different anchoring groups, we have obtained a much reduced attenuation factor. In addition, for some specific anchoring groups, it is even possible to increase the overall conductance of the molecular junction. By considering platinum instead of gold, yielding the same effect, we have deduced that the attenuation factor reduction can be attributed to symmetry breaking induced by the graphene electrode in the molecular junction. Indeed, at one side, the molecule is coupled covalently to the electrode, whereas it is weakly coupled (through van der Waals interactions) at the other side. Another interesting case to consider is a full asymmetric junction, with different anchoring groups and different electrodes at each side. As a result, the most conductive anchoring group, similarly to what happens at the macroscopic scale, mainly drives the conductance. Finally, hybrid junctions with conjugated molecular wires have been studied as well. A much lower attenuation factor could have been expected since these molecules are very conductive, unfortunately, due to bad coupling to the electrodes, the electronic transport properties remain unchanged, independently of the electrodes or the anchoring groups.

To conclude, despite the recent advances in the measurement techniques as well as in the theoretical and computational approaches, a full understanding of the electronic transport mechanism in molecular junctions has not been reached yet. In particular, the theoretical determination of the attenuation factor of the current remains a difficult task due to the lack of universal model to treat this problem. As we have seen, the Simmons model applies in very specific cases where the HOMO level is close to the Fermi level, as in the case of alkanedithiol in hybrid gold–graphene junctions. For other anchoring groups, the difference in symmetry couplings strongly modifies the energetic position of the molecular states, and the Simmons model does not apply anymore. Other theoretical approaches do not apply either since they consider infinite molecular chains and no effect of anchoring groups or electrodes. Therefore, the attenuation factors for these systems are theoretically deduced from this first case in comparison with the standard symmetric junctions, which are well documented in the literature. The attenuation factors of these hybrid junctions are all very similar, due to compensation effects between coupling and conduction in the molecular backbone and coupling to the electrodes. Moreover, we have also observed that the complex bandstructure approach may apply to aromatic molecular chains since the attenuation factors do not differ from the one determined in symmetric junctions. Consequently, a challenging task and an ideal perspective for theoretical approaches in determining attenuation factors in molecular junctions would be to consider, in the meantime, the intrinsic properties of the molecular backbone and the coupling to the electrodes through the anchoring groups that determines the positions of the molecular levels with respect to the Fermi level.

Funding: This research received no external funding.

Conflicts of Interest: The authors declare no conflict of interest.

References

1. Cuevas, J.C.; Scheer, E. *Molecular Electronics: An Introduction to Theory and Experiment*; World Scientific: Singapore, 2010.
2. Wang, W.; Lee, T.; Reed, M.A. Mechanism of electron conduction in self-assembled alkanethiol monolayer devices. *Phys. Rev. B* **2003**, *68*, 035416. [CrossRef]
3. Sze, S.M. *Physics of Semiconductor Devices*, 2nd ed.; Wiley: New York, NY, USA, 1981.
4. Hines, T.; Diez-Perez, I.; Hihath, J.; Liu, H.; Wang, Z.S.; Zhao, J.; Zhou, G.; Müllen, K.; Tao, N. Transition from Tunneling to Hopping in Single Molecular Junctions by Measuring Length and Temperature Dependence. *J. Am. Chem. Soc.* **2010**, *132*, 11658. [CrossRef] [PubMed]
5. Segal, D.; Nitzan, A.; Davis, W.B.; Wasielewski, M.R.; Ratner, M.A. Electron Transfer Rates in Bridged Molecular Systems 2. A Steady-State Analysis of Coherent Tunneling and Thermal Transitions. *J. Phys. Chem. B* **2000**, *104*, 3817. [CrossRef]
6. Nitzan, A. The relationship between electron transfer rate and molecular conduction 2. The sequential hopping case. *Isr. J. Chem.* **2002**, *42*, 163. [CrossRef]
7. Yan, H.; Bergren, A.J.; McCreery, R.; Luisa Della Rocca, M.; Martin, P.; Lafarge, P.H.; Lacroix, J.C. Activationless charge transport across 4.5 to 22 nm in molecular electronic junctions. *Proc. Natl. Acad. Sci. USA* **2013**, *110*, 5326. [CrossRef]
8. Choi, S.H.; Kim, B.; Frisbie, C.D. Electrical Resistance of Long Conjugated Molecular Wires. *Science* **2008**, *320*, 1482. [CrossRef]
9. Von Hippel, A.R. Molecular engineering. *Science* **1956**, *123*, 315. [CrossRef]
10. Kamenetska, M.; Koentopp, M.; Whalley, A.C.; Park, Y.S.; Steigerwald, M.L.; Nuckolls, C.; Hybertsen, M.S.; Venkataraman, L. Formation and Evolution of Single-Molecule Junctions. *Phys. Rev. Lett.* **2009**, *102*, 126803. [CrossRef]
11. Li, C.; Pobelov, I.; Wandlowski, T.; Bagrets, A.; Arnold, A.; Evers, F. Charge Transport in Single Au | Alkanedithiol | Au Junctions: Coordination Geometries and Conformational Degrees of Freedom. *J. Am. Chem. Soc.* **2008**, *130*, 318. [CrossRef]
12. Li, X.; He, J.; Hihath, J.; Xu, B.; Lindsay, S.M.; Tao, N. Conductance of Single Alkanedithiols: Conduction Mechanism and Effect of Molecule-Electrode Contacts. *J. Am. Chem. Soc.* **2006**, *128*, 2138. [CrossRef]

13. Klausen, R.S.; Widawsky, J.R.; Steigerwald, M.L.; Venkataraman, L.; Nuckolls, C. Conductive Molecular Silicon. *J. Am. Chem. Soc.* **2012**, *134*, 4541. [CrossRef] [PubMed]

14. Su, T.A.; Li, H.; Zhang, V.; Neupane, M.; Batra, A.; Klausen, R.S.; Kumar, B.; Steigerwald, M.L.; Venkataraman, L.; Nuckolls, C. Single-Molecule Conductance in Atomically Precise Germanium Wires. *J. Am. Chem. Soc.* **2015**, *137*, 12400. [CrossRef] [PubMed]

15. Li, H.; Garner, M.H.; Su, T.A.; Jensen, A.; Inkpen, M.S.; Steigerwald, M.L.; Venkataraman, L.; Solomon, G.C.; Nuckolls, C. Extreme Conductance Suppression in Molecular Siloxanes. *J. Am. Chem. Soc.* **2017**, *139*, 10212. [CrossRef] [PubMed]

16. Bonifas, A.P.; McCreery, R.L. 'Soft' Au, Pt and Cu contacts for molecular junctions through surface-diffusion-mediated deposition. *Nat. Nanotechnol.* **2010**, *5*, 612. [CrossRef] [PubMed]

17. He, J.; Chen, F.; Li, J.; Sankey, O.F.; Terazono, Y.; Herrero, C.; Gust, D.; Moore, T.A.; Moore, A.L.; Lindsay, S.M. Electronic Decay Constant of Carotenoid Polyenes from Single-Molecule Measurements. *J. Am. Chem. Soc.* **2005**, *127*, 1384. [CrossRef] [PubMed]

18. Capozzi, B.; Dell, E.J.; Berkelbach, T.C.; Reichman, D.R.; Venkataraman, L.; Campos, L.M. Length-Dependent Conductance of Oligothiophenes. *J. Am. Chem. Soc.* **2014**, *136*, 10486. [CrossRef]

19. Gunasekaran, S.; Hernangomez-Perez, D.; Davydenko, I.; Marder, S.; Evers, F.; Venkataraman, L. Near Length-Independent Conductance in Polymethine Molecular Wires. *Nano Lett.* **2018**, *18*, 6387. [CrossRef]

20. Sedghi, G.; Garcia-Suarez, V.M.; Esdaile, L.J.; Anderson, H.L.; Lambert, C.J.; Martin, S.; Bethell, D.; Higgins, S.J.; Elliott, M.; Bennett, N.; et al. Long-range electron tunnelling in oligo-porphyrin molecular wires. *Nat. Nano* **2011**, *6*, 517. [CrossRef]

21. Leary, E.; Limburg, B.; Alanazy, A.; Sangtarash, S.; Grace, I.; Swada, K.; Esdaile, L.J.; Noori, M.; Teresa Gonzalez, M.; Rubio-Bollinger, G.; et al. Bias-Driven Conductance Increase with Length in Porphyrin Tapes. *J. Am. Chem. Soc.* **2018**, *140*, 12877. [CrossRef]

22. Brooke, C.; Vezzoli, A.; Higgins, S.J.; Zotti, L.A.; Palacios, J.J.; Nichols, R.J. Resonant transport and electrostatic effects in single-molecule electrical junctions. *Phys. Rev. B* **2015**, *91*, 195438. [CrossRef]

23. Sayed, S.Y.; Fereiro, J.A.; Yan, H.; McCreery, R.L.; Bergren, A.J. Charge transport in molecular electronic junctions: Compression of the molecular tunnel barrier in the strong coupling regime. *Proc. Natl. Acad. Sci. USA* **2012**, *109*, 11498. [CrossRef] [PubMed]

24. Fereiro, J.A.; McCreery, R.L.; Bergren, A.J. Direct Optical Determination of Interfacial Transport Barriers in Molecular Tunnel Junctions. *J. Am. Chem. Soc.* **2013**, *135*, 9584. [CrossRef] [PubMed]

25. Martin, S.; Manrique, D.Z.; Garcia-Suarez, V.M.; Haiss, W.; Higgins, S.J.; Lambert, C.J.; Nichols, R.J. Adverse effects of asymmetric contacts on single molecule conductances of $HS(CH_2)n$ COOH in nanoelectrical junctions. *Nanotechnology* **2009**, *20*, 125203. [CrossRef] [PubMed]

26. Wang, K.; Zhou, J.; Hamill, J.M.; Xu, B. Measurement and understanding of single-molecule break junction rectification caused by asymmetric contacts. *J. Chem. Phys.* **2014**, *141*, 054712. [CrossRef] [PubMed]

27. Evers, F.; Korytar, R.; Tewari, S.; van Ruitenbeek, J.M. Advances and challenges in single-molecule electron transport. *Rev. Mod. Phys.* **2020**, *92*, 035001. [CrossRef]

28. Simmons, J.G. Generalized Formula for the Electric Tunnel Effect between Similar Electrodes Separated by a Thin Insulating Film. *J. Appl. Phys.* **1963**, *34*, 1793. [CrossRef]

29. Nitzan, A. Electron Transmission Through Molecules and Molecular Interfaces. *Ann. Rev. Phys. Chem.* **2001**, *52*, 681. [CrossRef]

30. Tomfohr, J.K.; Sankey, O.F. Complex band structure, decay lengths, and Fermi level alignment in simple molecular electronic systems. *Phys. Rev. B* **2002**, *65*, 245105. [CrossRef]

31. Picaud, F.; Smogunov, A.; Corso, A.D.; Tosatti, E. Complex band structures and decay length in polyethylene chains. *J. Phys. Condens. Matter.* **2003**, *15*, 3731. [CrossRef]

32. Desjonqueres, M.C.; Spanjaard, D. Concepts in surface physics. In *Springer Series in Surface Science*; Springer: Berlin, Germany, 1993; Volume 30.

33. Kiguchi, M.; Kaneko, S. Single Molecule Bridging Between Metal Electrodes. *Phys. Chem. Chem. Phys.* **2013**, *15*, 2253–2267. [CrossRef]

34. Quek, S.Y.; Choi, H.J.; Louie, S.G.; Neaton, J.B. Length Dependence of Conductance in Aromatic Single-Molecule Junctions. *Nano Lett.* **2009**, *9*, 3949–3953. [CrossRef] [PubMed]